STOCHASTIC AUTOMATA

Constructive Theory

A. A. Lorenz

STOCHASTIC AUTOMATA
Constructive Theory

Translated from Russian by D. Louvish

A HALSTED PRESS BOOK

JOHN WILEY & SONS
New York · Toronto

ISRAEL PROGRAM FOR SCIENTIFIC TRANSLATIONS
Jerusalem · London

Sole distributors for the Western Hemisphere and Japan

HALSTED PRESS, a division of
JOHN WILEY & SONS, INC., NEW YORK

Library of Congress Cataloging in Publication Data

Lorents, Aĭvar Arvidovich.
Stochastic automata, constructive theory.

Translation of *Elementy konstruktivnoĭ teorii veroi͡atnostnykh avtomatov.*
"A Halsted Press book."
Bibliography: p.
1. Sequential machine theory. 2. Constructive mathematics. I. Title.
QA267.5.S4L6513 629.8'91 74-8184
ISBN 0-470-54750-2

Distributors for the U.K., Europe, Africa and the Middle East

JOHN WILEY & SONS, LTD., CHICHESTER

Distributed in the rest of the world by

KETER PUBLISHING HOUSE JERUSALEM LTD.
ISBN 0 7065 1379 7
IPST cat. no. 22098

This book is a translation from Russian of
ELEMENTY KONSTRUKTIVNOI TEORII
VEROYATNOSTNYKH AVTOMATOV
Izdatel'stvo "Zinatne"
Riga, 1972

Printed in Israel

CONTENTS

PREFACE

This book summarizes the author's research on the constructive theory of probability and stochastic automata, carried out from 1965 at the Institute of Electronics and Computer Technology of the Academy of Sciences of the Latvian SSR. No previous knowledge is demanded of the reader other than a general familiarity with the elementary theory of algorithms.* An exception to this rule are the notes appended to each chapter, which are aimed primarily at those proficient in the theory of stochastic automata or those desirous of acquiring a more detailed acquaintance with the subject.

This book may cause certain difficulties for the reader not familiar with the ideas of constructive mathematics. We have therefore seen fit to include a small appendix on logic, explaining the basic principles of constructive mathematics; moreover, in some cases we present more detailed proofs than actually necessary. This should, we believe, enable the reader to gain some practice in reasoning in accordance with constructivistic principles.

A few words should be said on the very need for this book, which is a rather unusual addition to the literature on automata theory. There are two main reasons. First, we would like to preserve in the theory of stochastic automata the clear-cut constructive lines that characterize the theory of finite deterministic automata. Second, we wish to illustrate the ability of constructive mathematics to cope with a field which is quite close to the practical world. We hope that the reader will regard our efforts with understanding and without prejudice. The author is indebted to I. A. Metra and K. M. Podnieks for their critical remarks, which contributed much to the book, and also to M. L. Vevere for her active assistance in preparing the manuscript for the press.

A. A. Lorenz

* [The Western reader may find a full account of Markov's theory of algorithms in the English translation of his book (1954) and also in A.I. Mal'tsev's "Algorithms and Recursive Functions" (Moscow, Nauka, 1965), an English translation of which has been published by Wolters-Noordhoff (Groningen, 1970). — Trans.]

Introduction

THE NATURE OF STOCHASTIC AUTOMATA

§1. PHYSICAL MEANING OF THE MODELS

In this book we shall examine mathematical models of machines or physical devices of the following types: deterministic and stochastic finite automata; deterministic data-processing devices with potentially unlimited memory, whose inputs accept a potentially infinite tape on which information has been recorded by some algorithmic or stochastic procedure.

By a finite automaton (in the intuitive sense of the word) we mean a physical device capable of finitely many states $s_1, s_2, \ldots, s_n$ and changing these states (passing from state s_i to state s_j) only upon reception of an external signal from a finite set $\sigma_1, \sigma_2, \ldots, \sigma_m$. It is assumed that the external signals σ_k and σ_l cannot overlap, i. e., σ_l may begin only after σ_k has ended. If the law governing changes of state is dynamic in nature, i. e., the next state is uniquely determined by the external signal, the device and its state when the signal is applied, the automaton is said to be deterministic. If on the contrary this law is statistical in nature and the next state is determined only with a certain probability, we are dealing with a stochastic automaton.

Whether one regards a given physical device as deterministic or not depends largely on one's viewpoint. The same is true as regards finite stochastic automata. For example, an electronic computer may be viewed as a good instance of a finite deterministic automaton, provided one ignores many factors which nevertheless exert a considerable influence on its performance: temperature of vacuum tubes or transistors, fluctuations in power supply, and so on. On the other hand, the computer is also an example of a stochastic automaton, since its transitions from state to state often involve random malfunctions. In the final analysis, one should not forget that dynamic and statistical laws governing the performance of a physical device are only idealizations of reality. Consequently, when one speaks of a given physical device as exemplifying a deterministic or stochastic finite automaton, one is thereby adopting a certain idealization and ignoring some factor or factors. One is never in a position exhaustively to describe all the features, properties and factors being ignored. Thus, in speaking of a deterministic (or stochastic) finite automaton one envisages a physical device possessing properties which, to a certain degree, agree with intuitive ideas of what it means to say that it has finitely many internal states and external signals, and with one's conception of the strict determinism or statistical nature of its operation. Note that our notion of a deterministic finite automaton permits random fluctuations in the duration of the transition process. A finite automaton may be deterministic even when the time taken to change from state to state is governed by some statistical law.

As mentioned, our book will also discuss mathematical models of two other types of machine: deterministic and stochastic infinite machines. In point of fact, one can discuss infinite machines as physical devices only in an idealized sense. The idealization here is qualitatively different from that involved in the concept of a finite automaton. To our mind, the term "infinite machines" is appropriate for physical devices capable of executing a finite number of distinct elementary operations $O_1, O_2, \ldots, O_m$, but having an unlimited memory. The devices themselves function, by assumption, in a strictly deterministic manner: their performance follows a set program and does not exhibit statistical features unless the external signals are applied in a random manner. The data to be processed by the machine should be recorded as a sequence of symbols on tape. The machine produces another sequence of symbols, also on a tape. Both input and output data are represented by symbols from some finite list. We are disregarding the physical nature of the symbols, the material of which the tape is made, and the technical details of the actual physical device. All we are assuming is that the input and output tapes are potentially infinite in length. An infinite machine of this type will be called deterministic if the input data are recorded in accordance with a well-defined algorithm. An infinite machine will be stochastic if the symbols in the sequence representing the input data are only statistically defined, i. e., they are governed by some preassigned probability distribution.

The prototype of an infinite (deterministic or stochastic) machine is a digital computer, provided one ignores factors such as technical limitations on the design of absolutely reliable electronic devices and infinitely long punched tapes. Neither should one forget that no actual procedure is really an absolutely accurate realization of a stochastic law.

Reviewing the above discussion, we conclude that the mathematical models to be examined below are idealizations, in the sense that they may be termed "models" only subject to certain reservations, and the degree of idealization increases in proportion to the amount of transfinite elements in the models. It is worth pointing out, however, that apart from their traditional interpretation the mathematical models to be considered admit another interpretation, based on learning processes. We now know of other real processes which are amenable to description by these models.

§2. FINITE AUTOMATA

The first significant research on mathematical models of finite deterministic automata (FDA) dates back to Moore (1956), Kleene (1956) and Medvedev (1956). In Kleene's paper we find the beginnings of structural synthesis, while Moore and Medvedev devote all their attention to questions of analysis. Kleene (followed by Medvedev) presents a highly detailed analysis of the structure of events representable by FDA, and delineates the range of application of these automata. Moore's paper is interesting for its profound investigation of decoding problems. The only fresh ideas to appear in this field since then are those of Barzdin' (1969b).

Mathematical models of FDA have been studied by many authors. Here we shall only mention a few of the relevant publications.

Many important questions of structural synthesis have been studied by Caldwell (1958), Huffman (1954, 1957), and Gavrilov (1950).

The range of problems treated in the investigation of mathematical models of FDA was later considerably enlarged by McNaughton, Büchi, Glushkov, Trakhtenbrot and Barzdin'. The reader may find these results in Glushkov (1962), Kobrinskii and Trakhtenbrot (1959), Moisil (1959), Trakhtenbrot and Barzdin' (1970).

The first attempts to investigate finite stochastic automata (FSA) using mathematical models were almost contemporary with the first studies of FDA. The reasons motivating those who tackled this subject were different. For example, von Neumann (1956), in his celebrated lectures on stochastic logic at Caltech (1952), was primarily interested in questions of reliability. Consequently, his paper is permeated by a negative interpretation of the statistical phenomena observed in the operation of finite automata; in other words, randomness is viewed by von Neumann as an effect to be eliminated, something to be fought. Ashby (1956) was guided by completely different motives. He introduced his concept of a Markov machine as a mathematical model of the functioning of certain real systems directly related to biology. In so doing, he stressed the positive aspects of the statistical phenomena observed there, which in his opinion bear the primary responsibility for the simplicity of certain control mechanisms. This idea was later given a mathematical interpretation by Rabin (1963a) and Kovalenko (1965).*

Concurrently with Ashby, Medvedev (1956) proposed a mathematical model of FSA. His intention was simply to present a natural generalization of the mathematical model of FDA, specifically, of Moore automata. It appears that Medvedev did not think his model of interest, since he devoted little attention to it. Medvedev's model was essentially the same as Ashby's Markov machine. Both, however, are prototypes of the modern classical definition of a mathematical FSA, that due to Carlyle (1963) and Rabin (1963). These papers marked the beginnings of the true mathematical theory of FSA. Prior to their publication we find either research only loosely connected with the subject (von Neumann, 1956; Bush and Mosteller, 1955), or the definition alone (Ashby, 1956; Medvedev, 1956). Carlyle's main contribution to the theory of FSA was the first penetrating investigation of the problem of minimizing the number of internal states of an FSA. This problem was later considered by Bacon (1964a) and Starke (1969a, b). As for Rabin's papers (1963a, b), these deal with a variety of topics. Rabin studies the structure of events representable by FSA, stability of FSA, i. e., invariance of output under small variations of the transition probabilities, and, finally, the problem of state minimization or simplification of automata; his approach is via comparison of stochastic and deterministic automata. Some of these problems have been developed further by Paz, Turakainen, Salomaa, Bukharaev, Kochkarev and Starke.

The work of Bush and Mosteller, Tsetlin, Varshavskii, Shreider, etc., ** departs somewhat from the main trend of the classical theory. These

* In another connection (operation of a telephone exchange) the same idea of the positive value of statistical regularities was suggested and worked out mathematically by Feller (1950).

** See, e.g., the extensive bibliography in Bush and Mosteller (1955).

authors consider mathematical models of learning and also learning processes employing the models. For various reasons, however, their models have much in common with mathematical models of FSA. The problems that they solve, however, are quite different. One reason is of course the specific nature of the subject matter. Without going into details, we mention that Tsetlin, for example, is concerned mainly with mean values of the number of correct responses and with questions of behavior optimization; he does not refer at all to the type of problem considered by Carlyle and Rabin.

Lofgren (1965) considers a completely different type of mathematical model for FSA. This model differs from the traditional ones in its explicit structural character. Underlying Lofgren's arguments are both considerations of mathematical generality and motives rooted in biology and quantum mechanics. His starting point is the deterministic logical net of von Neumann-Church.

In the Soviet Union, structural models of FSA have been studied by Makarevich, who has proposed the foundations of an asynchronous theory of structural synthesis for FSA.

§3. MODELS OF INFINITE MACHINES

A history of mathematical models of infinite deterministic machines (IDMs) would be practically the same as a history of the theory of algorithms. Suffice it to mention, therefore, that the foundations of this area of research were laid by Turing (1936 — 1937) and, to some extent, Post (1936). What Turing machines can do is now common knowledge, though there are still plenty of unsolved problems. There is a tremendous literature on the subject; a full survey would require far too much space and would moreover be rather remote from the main subject of our book. The only publications of some relevance for our purposes are those considering what is known as the complexity of algorithms. Many authors have made important contributions to this aspect of the theory: Markov, Trakhtenbrot, Blum, Barzdin', Rabin, Tseitin, and others. The notion of complexity is interpreted in this context as either the complexity of a computation process (e. g., in computation of some arithmetical function) or the complexity of the computation program itself. For each type of complexity there is a specific quantitative measure.

Mathematical models for infinite stochastic machines (ISMs) were first considered by de Leeuw et al. (1956), who based themselves on the theory of Turing machines. We have in view theorems on the undecidability of certain general problems. In order to show that ISMs are essentially more powerful than IDMs, an attempt was made to ascertain in terms of mathematical models what the former can do. However, these attempts did not justify the hopes invested in the early models and in the first endeavors to state the problem rigorously. The positive results of de Leeuw et al. cannot be taken seriously, in view of their highly complicated realization. The authors themselves were well aware of this. Barzdin' (1969), taking de Leeuw et al. as his point of departure, modified their statement of the problem and unexpectedly reached a positive result of a more promising nature: he showed that ISMs can enumerate nonenumerable sets. There is nevertheless a price to pay: one cannot predict the law by which the set is defined.

Pertinent to this range of problems is the work of Agafonov and (apparently) Irving, as indicated by the latter's conjecture that any stochastically computable function $\varphi(x)$ is recursive. However, we have been unable to ascertain the precise meaning of the term "stochastically computable function" from the abstract of Irving's dissertation. We are therefore still unable to answer the question as to whether this extension of Church's thesis has a bearing on the results of de Leeuw et al.

§4. HISTORICAL ORIGINS OF THE CONSTRUCTIVE THEORY OF STOCHASTIC AUTOMATA

The beginnings of the constructive approach to the analysis of stochastic machines may be found in the above-mentioned paper of de Leeuw et al. (1956). Although this work is on the whole classical in tendency, it includes constructive mathematical models of ISMs and a constructive analysis of their properties. Constructive mathematical models for FSA in a general setting were first discussed by the present author, although the need for research of this kind was already felt by Rabin. In regard to FDA and IDMs, the existing mathematical models were and still are constructive, irrespective of the philosophies of the various authors. The results obtained in these fields are in many cases also constructive. The only palpable break between the constructive and classical theories of automata is evident with regard to mathematical models of FSA or ISMs. In order to preserve constructive principles in the investigation of such models, the author has undertaken to lay the foundations for a constructive theory of FSA and ISMs. The same considerations that impelled de Leeuw, Moore, Shannon, Shapiro and Rabin to suggest constructive models have played a major role here. These considerations essentially reduce to the statement that certain questions of the theory of FSA and ISMs cannot be formulated correctly without appeal to constructive mathematical models.

This concludes our historical survey of the development of the theory of stochastic automata. Brief historical notes will be given at the end of each chapter.

Chapter I

BASIC CONCEPTS OF CONSTRUCTIVE MATHEMATICS

§1. CONSTRUCTIVE REAL NUMBERS

We shall use the following fixed alphabets:

$$A_0 = \{0, |\};$$
$$A_1 = \{0, |, -\};$$
$$A_2 = \{0, |, -, /\};$$
$$A_3 = \{0, |, -, /, \Diamond\}.$$

We shall view the natural numbers as words over the alphabet A_0, defined by the following generating rule: (1) 0 is a natural number (zero); (2) if a word P is a natural number, then the word $P|$ is also a natural number (the successor of P). Natural numbers will be called words of type N. Integers (words of type I) are defined as words over the alphabet A_1, defined as follows: for every n_1, the words n_1 and $-n_1|$ are integers ($-n_1|$ is the negative number of absolute value $n_1|$). Rational numbers (words of type R) are words over A_2 defined as follows: (1) every integer is a rational number; (2) for any n_1 and c_1 (where n_1 is a natural number and c_1 an integer), the word $c_1/n_1|$ is a rational number (the fraction with numerator c_1 and denominator $n_1|$).

The operations of addition, subtraction, multiplication and division in the field of rational numbers (words of type R) are defined by suitable algorithms, which agree with the familiar rules of addition, subtraction, multiplication and division of rational numbers in the ordinary decimal notation. For example, addition of natural numbers m, n is performed with the help of a normal algorithm $\mathfrak{A}_+$:

$$\begin{cases} |+0| \to ||+0; \\ +0 \to \quad , \end{cases}$$

and subtraction via an algorithm $\mathfrak{A}_-$:

$$\begin{cases} |-0| \to -0; \\ |-0 \to |; \\ 0-0| \to -0|; \\ 0-0 \to 0. \end{cases}$$

Applying these algorithms to the words $(m+n)$ and $(m-n)$, respectively, we obtain the sum and difference of numbers m and n. Of course, the translation of our abstract logical form of natural numbers, integers and rationals to the usual notation and back may be accomplished by using suitable algorithms. For convenience, therefore, we shall for the most part work with rational numbers in their usual notation, i. e., quotients of two natural numbers in decimal notation.

We shall also assume that algorithms are available for the most important arithmetical functions, and also the functions $|x|$ and $E(x)$ — absolute value and integral part of a number x.

The relations $=$ and $<$ for numbers of type R will be introduced by the following definitions.

If m and n are integers, the notation $m=n$ will mean that $m \overline{\overline{\underline{\circ}}} n$ [i. e., m and n are "graphically equal" — see Appendix]. If m is an integer and $r \overline{\overline{\underline{\circ}}} n/n_1|$, then $m=r$ means that $mn_1|=n$.

In the case of rational numbers $r \overline{\overline{\underline{\circ}}} c_1/m|$ and $s \overline{\overline{\underline{\circ}}} c_2/n|$, the notation $r=s$ will mean

$$c_1 n| = c_2 m|.$$

A rational number r is said to be positive if $r \neq 0$ and $|r|=r$.

We shall say that a rational number r is smaller than a rational number s, writing $r<s$, if $(s-r)$ is a positive rational number. The relation $>$ is defined by the formula

$$(r>s) \leftrightarrow (s<r).$$

Our real numbers will be introduced via the duplex concept, which is defined as follows. An algorithm $\mathfrak{B}$ of type (N → N)* will be called a convergence regulator for an algorithm $\mathfrak{A}$ of type (N → R) if for any n_1, n_2, n_3 such that $n_2 \geqslant \mathfrak{B}(n_1)$ and $n_3 \geqslant \mathfrak{B}(n_1)$ we have

$$|\mathfrak{A}(n_2) - \mathfrak{A}(n_3)| < 2^{-n_1}.$$

The reader should note that algorithms are notated as words over the alphabet A_0. A duplex (word of type g) is defined thus: (1) any rational number; (2) any word of the form $\mathfrak{A}_1 \diamond \mathfrak{A}_2$, where $\mathfrak{A}_1$ is the notation for some algorithm $\mathfrak{A}$ of type (N → R) in a standard extension of the alphabet A_2, and $\mathfrak{A}_2$ the notation for some algorithm $\mathfrak{B}$ of type (N → N) in a standard extension of the alphabet A_0, which is a convergence regulator. Duplexes are words over the alphabet A_3.

The duplex concept is analogous to the classical definition of real numbers through mutually convergent sequences of rational numbers (in our approach the role of the sequences of rational numbers is played by algorithms of type (N → R)); in a sense, it is a variant of Specker's computable real number (Specker, 1949). In the sequel, however, we shall use the term constructive real numbers not only for such words $\mathfrak{A}_1 \diamond \mathfrak{A}_2$ over the alphabet A_3 but also for ordered pairs of normalized

* An algorithm of type (N → N) is an algorithm which, when applied to any word of type N, produces a word of type N. The definition of an algorithm of type (N → R) is analogous.

algorithms $\mathfrak{A}$, $\mathfrak{B}$ (or ordered pairs of normalized systems of word-transformation rules), satisfying the following conditions:

(1) The algorithms $\mathfrak{A}$ and $\mathfrak{B}$ are applicable to any natural number n.
(2) The values of the algorithm $\mathfrak{A}$ (denoted by $\mathfrak{A}(n)$) are rational numbers.
(3) The values of the algorithm $\mathfrak{B}$ (denoted by $\mathfrak{B}(n)$) are natural numbers.
(4) For any natural numbers m and p,

$$|\mathfrak{A}(\mathfrak{B}(m)) - \mathfrak{A}(\mathfrak{B}(m)+p)| < \frac{1}{f(m)},$$

where $f(m)$ is a positive increasing arithmetical function.

We now define the basic operations over duplexes and the basic relations between them. Let d_1 and d_2 be two given duplexes, where $d_1 \boxeq \mathfrak{A}_1 \diamond \mathfrak{A}_2$ and d_2 is a word of type R. Then the sum d_1+d_2 is defined as a new duplex $d \boxeq \mathfrak{B}_1 \diamond \mathfrak{B}_2$. If we let $\mathfrak{A}$, $\mathfrak{A}'$, $\mathfrak{B}$, $\mathfrak{B}'$ denote the algorithms whose notations are the words $\mathfrak{A}_1$, $\mathfrak{A}_2$, $\mathfrak{B}_1$, $\mathfrak{B}_2$, respectively, the operation of the algorithms $\mathfrak{B}$ and $\mathfrak{B}'$ is defined as follows:

$$\mathfrak{B}(n) = \mathfrak{A}(n) + d_2;$$
$$\mathfrak{B}'(n) = \mathfrak{A}'(n).$$

The difference (d_1-d_2) is defined similarly, the operation of the algorithm $\mathfrak{B}$ being defined by

$$\mathfrak{B}(n) = \mathfrak{A}(n) - d_2.$$

The sum (d_2+d_1) and difference (d_2-d_1) are defined symmetrically.

For the product d_1d_2, the algorithms $\mathfrak{B}$ and $\mathfrak{B}'$ are defined thus:

$$\mathfrak{B}(n) = \mathfrak{A}(n)\, d_2;$$
$$\mathfrak{B}'(n) = \mathfrak{A}'(n+E(|d_2|)).$$

For division $\left(\frac{d_1}{d_2}\right)$, the number d_2 must be distinct from zero. Thus definition of the quotient $\frac{d_1}{d_2}$ is equivalent to definition of the product $\left(d_1 \cdot \frac{1}{d_2}\right)$.

We now define the sum, difference and product of duplexes d_1 and d_2 when $d_2 \boxeq \widehat{\mathfrak{A}}_1 \diamond \widehat{\mathfrak{A}}_2$.

Let $\mathfrak{A}$, $\mathfrak{A}'$, $\widehat{\mathfrak{A}}$, $\widehat{\mathfrak{A}}'$ denote the algorithms whose notations are the words $\mathfrak{A}_1$, $\mathfrak{A}_2$, $\widehat{\mathfrak{A}}_1$, $\widehat{\mathfrak{A}}_2$, respectively. Then the sum (d_1+d_2) of d_1 and d_2 is the duplex d, $d \boxeq \mathfrak{B}_1 \diamond \mathfrak{B}_2$, defined by

$$\mathfrak{B}(n) = \mathfrak{A}(n) + \widehat{\mathfrak{A}}(n);$$
$$\mathfrak{B}'(n) = \max\,(\mathfrak{A}'(n+1),\ \widehat{\mathfrak{A}}'(n+1)).$$

The difference (d_1-d_2) is the duplex d defined by

$$\mathfrak{B}(n) = \mathfrak{A}(n) - \widehat{\mathfrak{A}}(n);$$
$$\mathfrak{B}'(n) = \max\,(\mathfrak{A}'(n+1),\ \widehat{\mathfrak{A}}'(n+1)).$$

The product d_1d_2 is the duplex d defined by

$$\mathfrak{B}(n)=\mathfrak{A}(n)\widehat{\mathfrak{A}}(n);$$
$$\mathfrak{B}'(n)=\max(\mathfrak{A}'(n+c),\ \widehat{\mathfrak{A}}'(n+c)),$$

where

$$c=\max\left(E\left(|\mathfrak{A}(\mathfrak{A}'(1))|+\frac{1}{2}\right),\ E\left(|\widehat{\mathfrak{A}}(\widehat{\mathfrak{A}}'(1))|+\frac{1}{2}\right)\right)+1.$$

We shall say that a duplex d, $d \eqcirc \mathfrak{A}_1 \diamond \mathfrak{A}_2$ vanishes (notation: $d=0$) if for every m there exists n such that whenever $k\geqslant n$

$$|\mathfrak{A}(k)|\leqslant\frac{1}{m}.$$

We can now define equality of any two duplexes: $d_1=d_2$ if $d_1-d_2=0$. Now consider a duplex d, $d \eqcirc \mathfrak{A}_1 \diamond \mathfrak{A}_2$. We denote by $|d|$ the duplex defined by the following pair of algorithms $\mathfrak{B}$ and $\mathfrak{B}'$:

$$\mathfrak{B}(n)=|\mathfrak{A}(n)|;$$
$$\mathfrak{B}'(n)=\mathfrak{A}'(n).$$

A duplex d is said to be positive (a positive constructive real number) if $|d|=d$ and $d\neq 0$.

We shall say that a duplex d_1 is smaller than a duplex d_2, writing $d_1<d_2$, if (d_2-d_1) is a positive constructive real number (CRN). The notation $d_1>d_2$ means that $d_2<d_1$.

If d, $d \eqcirc \mathfrak{A}_1 \diamond \mathfrak{A}_2$, is a positive duplex, then $d>0$. Indeed, we have $d-0=d$ and so $d-0$ is a positive CRN. Thus, by definition, $d>0$. A duplex d is said to be negative (a negative CRN) if $0-d$ is a positive CRN. Clearly, if d is a negative CRN then $d<0$.

Using the method of constructive choice, one can prove the following proposition:

If $d\neq 0$, there exist a natural number k and a positive rational number r such that

$$\forall n(|\mathfrak{A}(k+n)|\geqslant r). \tag{1}$$

It is noteworthy that this proposition is unprovable in intuitionistic mathematics, since intuitionsm rejects the principle of constructive choice.* Supposing that $d\neq 0$, we now define the quotient $\frac{1}{d}$. A duplex d', $d' \eqcirc \mathfrak{B}_1 \diamond \mathfrak{B}_2$, is called the quotient upon division of 1 by d if

$$\mathfrak{B}(n)=(\mathfrak{A}(n+k))^{-1};$$
$$\mathfrak{B}'(n)=\mathfrak{A}'(n+E(r^{-2})),$$

where k and r are numbers satisfying (1).

* See Markov (1962) and Heyting (1956).

We can now define the quotient $\frac{d_1}{d_2}$ upon division of d_1 by d_2, $d_2 \neq 0$, as the duplex $d_1 \cdot \frac{1}{d_2}$.

The CRNs form a complete separable metric space with respect to the metric

$$\varrho(d_1, d_2) = |d_1 - d_2|.$$

Whereas in classical mathematics the set of CRNs is enumerable, in the constructive system it is nonenumerable.

§2. CONSTRUCTIVE FUNCTIONS OF A REAL VARIABLE

Let d_1 and d_2 be CRNs such that $d_1 < d_2$. A CRN d such that $d_1 \not> d \not> d_2$ is said to belong to the interval $[d_1, d_2]$. If $d_1 < d < d_2$, we shall say that d belongs to the open interval (d_1, d_2).

A constructive function $F(z)$ of one real variable, defined in an interval $[d_1, d_2]$ (or (d_1, d_2)) is an algorithm $\mathfrak{F}$ satisfying the following conditions:

(1) $\mathfrak{F}$ is applicable to any CRN in the interval $[d_1, d_2]$ (or (d_1, d_2)), and it produces a CRN.

(2) When applied to equal CRNs, $\mathfrak{F}$ produces equal CRNs.

This definition of a constructive function is due to Markov (1958). Another definition of this concept, based on primitive recursion, was suggested by Goodstein (1945) and later studied by Specker (1949) and Goodstein himself (1961). Markov (1958) proved that a constructive function $F(z)$ defined in an interval $[d_1, d_2]$ has no points of discontinuity, and Tseitin (1962a) proved that it is continuous, i. e., one can find an algorithm $\mathfrak{A}$ with the following properties:

(1) $\mathfrak{A}$ is applicable to any word $m\square z$ (where m is a natural number and z a CRN in $[d_1, d_2]$).

(2) $\mathfrak{A}(m\square z)$ is a natural number.

(3) $\forall z' \left(\left(z - 2^{-\mathfrak{A}(m\square z)} < z' < z + 2^{-\mathfrak{A}(m\square z)} \right) \supset \left(F(z) - 2^{-m} < F(z') < F(z) + 2^{-m} \right) \right)$; z' is also in $[d_1, d_2]$.

However, the classical theorem that $F(z)$ is uniformly continuous cannot be proved constructively. The basic properties of constructive functions have been investigated by Tseitin (1962a, b) and Zaslavskii (1962).

Although a constructive function is by definition a normal algorithm, we shall not always represent it explicitly as a scheme of substitution formulas. For example, the polynomial $c_0 z^n + c_1 z^{n-1} + \ldots + c_n$ is also a constructive function, defined on any interval. This follows from the fact that this notation actually determines an algorithm which transforms any number z into another number z', equal numbers being transformed into equal numbers. The same is true of various other notations familiar from classical analysis. Appealing to the reader's intuition and ingenuity, we shall without further ado treat the expressions $\sqrt[n]{z}$, z^r, d^z, $\log_n z$, where n is a natural number greater than 1, r a rational number, and d a positive real number, as constructive functions defined in any interval $(0, d)$. In principle, all these functions may be defined rigorously, but we shall not do so; this should not lead to any misunderstandings. The situation is different for the functions $\min(z_1, z_2, \ldots, z_n)$ and $\max(z_1, z_2, \ldots, z_n)$ of several variables. These will therefore be defined in full rigor. It must first be noted that a constructive function of m real variables is defined as a natural generalization of a constructive function of one variable. To be precise: a function $F(z_1, z_2, \ldots, z_m)$ of m real variables defined in a closed m-dimensional cube $(d, d')^m$ is a normal algorithm $\mathfrak{F}$ such that:

(1) $\mathfrak{F}$ is applicable to any m-tuple of CRNs $(d_1, d_2, \ldots, d_m)$ such that $d \not> d_i \not> d'$, $i=1, 2, \ldots, m$.

(2) $\mathfrak{F}(d_1, d_2, \ldots, d_m)$ is a CRN.

(3) $((z_1, z_2, \ldots, z_m) = (z'_1, z'_2, \ldots, z'_m)) \supset (\mathfrak{F}(z_1, z_2, \ldots, z_m) = \mathfrak{F}(z'_1, z'_2, \ldots, z'_m))$

Equality of m-tuples means equality of the corresponding components:

$$\begin{cases} z_1 = z'_1; \\ z_2 = z'_2; \\ \ldots\ldots\ldots \\ \ldots\ldots\ldots \\ \ldots\ldots\ldots \\ z_m = z'_m. \end{cases}$$

We now proceed to define the function symbols $\min(z_1, z_2, \ldots, z_m)$ and $\max(z_1, z_2, \ldots, z_m)$. If $a_1, a_2, \ldots, a_m$ are rational numbers, $\min(a_1, a_2, \ldots, a_m)$ ($\max(a_1, a_2, \ldots, a_m)$) is the first of the numbers a_i, $i=1, 2, \ldots, m$, which is less than or equal to (greater than or equal to) all the others. Suppose now that $a_i \cong \mathfrak{A}_{1i} \diamond \mathfrak{A}_{2i}$. Denote the algorithms corresponding to the words $\mathfrak{A}_{1i}$ and $\mathfrak{A}_{2i}$ by $\mathfrak{A}_i$ and $\mathfrak{A}'_i$, respectively. Then a CRN d, $d \cong \mathfrak{B}_1 \diamond \mathfrak{B}_2$, is the maximum of $a_1, a_2, \ldots, a_m$ (notation: $\max(a_1, a_2, \ldots, a_m)$) if

$$\mathfrak{B}(n) = \max(\mathfrak{A}_1(n), \mathfrak{A}_2(n), \ldots, \mathfrak{A}_m(n));$$
$$\mathfrak{B}'(n) = \max(\mathfrak{A}'_1(n), \mathfrak{A}'_2(n), \ldots, \mathfrak{A}'_m(n)).$$

A CRN d is the minimum of $a_1, a_2, \ldots, a_m$ (notation: $\min(a_1, a_2, \ldots, a_m)$) if

$$\mathfrak{B}(n) = \min(\mathfrak{A}_1(n), \mathfrak{A}_2(n), \ldots, \mathfrak{A}_m(n));$$
$$\mathfrak{B}'(n) = \max(\mathfrak{A}'_1(n), \mathfrak{A}'_2(n), \ldots, \mathfrak{A}'_m(n)).$$

If some of the numbers $a_1, a_2, \ldots, a_m$ are rational CRNs, the corresponding numbers $\mathfrak{A}_i(n)$ in the formulas defining the algorithms $\mathfrak{B}$ and $\mathfrak{B}'$ should be replaced by a_i, and $\mathfrak{A}'_i(n)$ by unity.

The following properties of the functions min and max are easily proved.

Property 1. For any real numbers $a_1, a_2, \ldots, a_m$,

$$\max(a_1, a_2, \ldots, a_m) - \min(a_1, a_2, \ldots, a_m) = \\ = \max(a_i - a_j),\ 1 \leqslant i,\ j \leqslant m.$$

Property 2. Let $a_i, i=1, \ldots, m$, be real numbers and b a real number such that $a_i \not> b$ for all i; then

$$\max(a_1, a_2, \ldots, a_m) \not> b.$$

Examples of function expressions which cannot be accepted as constructive functions of one real variable are the classical functions sign(z) and $E(z)$. Both of these are functions only in the field of rational numbers.

§3. CONSTRUCTIVE SETS

The set concept is employed in constructive mathematics to denote constructive objects, or words, which represent conditions imposed upon constructive objects for which generation rules are available. A more precise definition is possible only in the context of some language of mathematical logic; an example is the language of Kleene (1952), provided that the basic constructive objects are natural numbers. Of course, the word "precision" in this context implies whatever level of mathematical rigor is required, at the present stage of development of constructive mathematics, in laying the basis for any special part of the theory. The above-mentioned approach to constructive sets, though necessary in logical analysis of certain properties of sets, nevertheless entails considerable difficulties if it is desired to employ the set concept in a creatively evolving mathematical theory. For example, surely no one would now question the possibility of defining decidable sets of natural numbers without going beyond the rigorous concept of a general recursive predicate. This essentially has no effect on the generality of the definition of the concept, but from a practical point of view it may sometimes be questionable, leading as it does quite often to superfluous technical difficulties. We must take this into account when defining a rigorous set concept for the theory now under construction; otherwise we shall find ourselves in a difficult position. A reasonable solution to the dilemma seems to be: formulate a rigorous definition of the set concept in terms of some formal logical language, using this as the basis for all proposed constructions. At the same time, however, agree to specify sets by other, different means, provided this does not entail an essential extension of the basic concept.

The stipulated conception of a set agrees in many respects with Kleene's views (Kleene, 1952, p. 274) on the definition of recursively enumerable sets and partial recursive functions.

Thus, suppose we are given some alphabet $B = \{\xi_1, \xi_2, \ldots, \xi_l\}$. The letters e, f and g (with or without subscripts) will denote words over B, the letter x being used as a variable for words over B.

An expression of the form

$$x \supseteq e \qquad (\alpha)$$

will be called an e l e m e n t a r y p r e d i c a t e f o r m u l a. We shall say that a word g satisfies (α) if the statement represented by the formula $g \supseteq e$ is true.

Let $\mathfrak{B}$ denote the code of the scheme of some normal algorithm over B, applicable to any word e over B and producing a real number; let c be some real number. The expressions

$$\mathfrak{B}(x)=c; \tag{β}$$
$$\mathfrak{B}(x)<c; \tag{γ}$$
$$\mathfrak{B}(x)>c \tag{δ}$$

will also be called elementary predicate formulas. Since the symbol $\mathfrak{B}$ stands for a well-defined algorithm, it is natural to let $\mathfrak{B}(e)$ denote the result of the algorithm when applied to a word e.

We shall say that e satisfies (β), (γ) or (δ) according as the statement $\mathfrak{B}(e)=c$, $\mathfrak{B}(e)<c$ or $\mathfrak{B}(e)>c$ is true.

The class M of predicate formulas is now defined as follows:

(1) Any elementary predicate formula is in M.

(2) If P and Q are predicate formulas in M, then $(P)\&(Q)$ and $(P)\vee(Q)$ are also in M.

(3) If a predicate formula P is in M, then $\neg P$ is also in M.

(4) A predicate formula may belong to M only by virtue of conditions (1), (2), (3).

Definition 1. We shall say that a predicate formula P defines a set $\{P\}$ of words over the alphabet B if and only if P is in M.

A word e will belong to the set $\{P\}$ (notation: $e\in\{P\}$) if e satisfies P.

The set $\{P\}$ is said to be nonempty if there exists a word e satisfying P, i. e., if the statement $\exists xP$ is true.

The notation $\forall x(x\in\{P\})$ means that every word e is in the set $\{P\}$, i. e., that the statement $\forall xP$ is true.

A set $\{P\}$ is said to be empty if the statement $\forall x\neg P$ is true, i. e., every word e is in the set $\{\neg P\}$.

An example of an empty set is that defined by the predicate formula

$$(x\overline{\underline{\circ}}e)\&\neg(x\overline{\underline{\circ}}e).$$

A set $\{P\}$ is called a subset of a set $\{Q\}$ if

$$\forall x((x\in\{P\})\supset(x\in\{Q\})).$$

The symbolic notation for this situation will be $\{P\}\subseteq\{Q\}$. The notation $\{P\}\approx\{Q\}$ will mean that $\{P\}\subseteq\{Q\}$ and $\{Q\}\subseteq\{P\}$.

We shall say that sets $\{P\}$ and $\{Q\}$ are equal if $\{P\}\approx\{Q\}$. It is readily seen that any two empty sets are equal. [We shall therefore speak loosely of "the" empty set. Similar abuses of language will be used without special mention.]

A set $\{Q\}$ is a complement of the set $\{P\}$ if $\{Q\}\approx\{\neg P\}$.

The complement of a set $\{P\}$ will be denoted by $\{\overline{P}\}$. It is clear that the sets $\{\overline{P}\}$ and $\{\overline{Q}\}$ are equal if $\{P\}$ and $\{Q\}$ are empty.

A set $\{R\}$ will be called an intersection of $\{P\}$ and $\{Q\}$ if it is equal to the set $\{P\&Q\}$. Our symbol for the intersection of sets $\{P\}$ and $\{Q\}$ will be $\{P\}\{Q\}$. It is easy to see that the intersection of a set $\{P\}$ and its complement is the empty set.

A set $\{R\}$ will be called a union of the sets $\{P\}$ and $\{Q\}$ if

$$\{R\} \approx \{P \vee Q\}.$$

The union of sets $\{P\}$ and $\{Q\}$ will be denoted by $\{P\} \cup \{Q\}$.

Definition 2. A set $\{P\}$ is said to be finite if it is equal to some set

$$\{(x \eqsim e_1) \vee (x \eqsim e_2) \vee \ldots \vee (x \eqsim e_m)\}^*$$

or to

$$\{(x \eqsim e) \& \neg (x \eqsim e)\}.$$

Clearly, once we have proved that a set $\{P\}$ is finite, we can determine the number of distinct (graphically different) words e in it. We denote this number by $\mu(\{P\})$, calling it the norm of the set $\{P\}$. According to this definition, the norm of an empty set $\{P\}$ is zero.

We now agree to identify graphically equal words e and e' belonging to a set $\{P\}$ and representing the same element of $\{P\}$; then $\mu(\{P\})$ (provided $\{P\}$ is finite) is the number of elements in the set $\{P\}$. The following properties of finite sets $\{P\}$ and $\{Q\}$ are easily verified

(1) *The union* $\{P\} \cup \{Q\}$ *is a finite set.*

(2) *The intersection* $\{P\}\{Q\}$ *is a finite set.*

(3) $(\{P\} \approx \{Q\}) \supset \mu(\{P\}) = \mu(\{Q\})$;

(4) $\mu(\{P \vee Q\}) \leqslant \mu(\{P\}) + \mu(\{Q\})$;

(5) $(\{P\} \approx \{P'\} \& \{Q\} \approx \{Q'\}) \supset (\{P\}\{Q\} \approx \{P'\}\{Q'\} \& \{P\} \cup \{Q\} \approx \{P'\} \cup \{Q'\} \& \& \{\overline{P}\}\{Q\} \approx \{\overline{P'}\}\{Q'\})$;

(6) $\{\overline{P}\}\{Q\}$ *is a finite set.*

As an illustration, we prove the last property.

Proof. If $\mu(\{P\}) = 0$, then $\{\overline{P}\}\{Q\} \approx \{Q\}$. Suppose now that $\mu(\{P\}) \neq 0$. If $\mu(\{Q\}) = 0$, we obtain

$$\{\overline{P}\}\{Q\} \approx \{Q\}.$$

Suppose, then, that $\mu(\{Q\}) \neq 0$. Since $\{P\}$ and $\{Q\}$ are finite sets, there exist sets $\{P'\}$ and $\{Q'\}$, represented respectively by predicate formulas

$$P' \equiv (x \eqsim e_1) \vee (x \eqsim e_2) \vee \ldots \vee (x \eqsim e_m)$$

and

$$Q' \equiv (x \eqsim e'_1) \vee (x \eqsim e'_2) \vee \ldots \vee (x \eqsim e'_n),$$

such that $\{P\} \approx \{P'\}$ and $\{Q\} \approx \{Q'\}$. By property (5), we have

$$\{\overline{P}\}\{Q\} \approx \{\overline{P'}\}\{Q'\} \approx \{(\neg (x \eqsim e_1) \& \neg (x \eqsim e_2) \& \ldots \& \neg (x \eqsim e_m)) \& ((x \eqsim e'_1) \vee (x \eqsim e'_2) \vee \ldots \vee (x \eqsim e'_n))\}.$$

* To simplify the notation we have omitted some of the parentheses.

The set $\{\overline{P}\}\{Q\}$ is empty if, for every e'_i, $1 \leqslant i \leqslant n$, we can choose e_j, $1 \leqslant j \leqslant m$, such that $e'_i \mathrel{\underline{\underline{\circ}}} e_j$. Otherwise,

$$\{\overline{P}\}\{Q\} \approx \{R\},$$

where

$$R \equiv (x \mathrel{\underline{\underline{\circ}}} e'_{i_1}) \vee \ldots \vee (x \mathrel{\underline{\underline{\circ}}} e'_{i_k});$$

$$e'_{i_j} \in \{Q\}; \quad e'_{i_j} \in \{\overline{P}\}; \quad j = 1, 2, \ldots, k.$$

Thus the set $\{\overline{P}\}\{Q\}$ is indeed finite.

§4. PRACTICAL CONVENTIONS AND GENERALIZATIONS

In the sequel we shall have to deal mainly with subsets of finite sets.* To abbreviate the notation, subsets $\{Q\}$ of a fixed finite set $\{P\}$ will frequently be represented by predicate formulas in one free variable x, which may possibly be meaningless for certain values of x. Such formulas are required to be meaningful only for words e in $\{P\}$. This convention of course adds nothing new to our system, for any such predicate formula may be redefined. We shall also adopt the following convention: a subset $\{Q \& P\}$ of a finite set $\{P\}$ will be represented only by the predicate formula Q, although $\{Q\}$ itself may not be a subset of $\{P\}$. This should not cause confusion, since we shall always indicate the basic finite set $\{P\}$.

In order to impart greater clarity to the set concept, we add a few remarks. Our definition of elementary predicate formulas appealed only to normal algorithms over the alphabet B, imposing no restrictions on the algorithms. Actually, however, it would suffice to use normal algorithms over a two-letter extension of the alphabet B. This is because any normal algorithm over B is equivalent with respect to B to some normal algorithm over $B \cup \{\alpha, \beta\}$ (see Markov, 1954, p. 145).**

A final remark: we stipulate that the representing formula of a set may be any predicate P in one free variable x, provided the class M contains an equivalent predicate formula Q. For example, one might consider a word $\xi_1\xi_2\xi_1$ and define with its aid a predicate $P(x)$, saying that $P(e)$ is true if and only if e occurs in the word $\xi_1\xi_2\xi_1$. Although $P(x)$ itself need not be in M, it may nevertheless be used to represent the set.

Regardless of the fact that the theory proposed below requires no essential extension of the set concept, we nevertheless present another definition. This definition may prove useful, say, if one desires to modify the definition of random event employed in this book.

Let x and y be variables for words over B, $\mathfrak{B}$ and $\mathfrak{B}'$ codes of schemes of any normal algorithms over B applicable to arbitrary words on B and producing real numbers. We define a class of relations R as follows:

* Note the following true statement: It is not true that every subset of a finite set is finite.

** This result was improved by Nagornyi (1953).

(1) Expressions of the form

$$\mathfrak{B}(x) = \mathfrak{B}'(y),$$
$$\mathfrak{B}(x) < \mathfrak{B}'(y)$$

and

$$\mathfrak{B}(x) > \mathfrak{B}'(y)$$

are relations of class R.

(2) If $P(x, y)$ and $Q(x, y)$ are relations of class R, then

$$(P(x, y)) \vee (Q(x, y))$$

and

$$(P(x, y)) \& (Q(x, y))$$

are also relations of class R.

(3) If $P(x, y)$ is a relation of class R, then $\neg(P(x, y))$ is also a relation of class R.

(4) An object may be a relation of class R only by virtue of conditions (1), (2), (3).

The class of elementary predicate formulas will include the expressions $x \equiv e$ and $\exists y P(x, y)$, where $P(x, y)$ is a relation of class R.

If we now confine the term "elementary predicate formula" to the sense just defined, the original definition of the class M becomes a definition of a new class of predicate formulas M'. The new set concept is obtained when M is replaced by M' in Definition 1.

NOTES

There are various definitions of the concept of CRN in the literature, but they are all equivalent. However, the very word "equivalent" is sometimes understood here in the classical sense; a good example of this situation is Uspenskii (1960). We, of course, are using this word in its constructive sense. In that case, however, the first definition of a CRN, due to Turing (1937), which was based on effective expression of a number in a well-defined number system, is not equivalent to the duplex definition. By contrast, the definition of CRN via a regular sequence of rational numbers is equivalent to the duplex concept or, in the terminology of Shanin (1962), these concepts model each other constructively.

A profound analysis of the properties of CRNs defined by primitive recursion has been made by Specker (1949).

Our definition of the set concept, as introduced in §3 and extended in §4, is not the broadest possible. A far more general definition is that based on extensive use of quantifiers and general recursive functions of several variables. For example, if $\varphi(x_1, x_2, \ldots, x_n, x)$ is a general recursive function and m a natural number, then the predicate

$$\forall x_1 \forall x_2 \exists x_3 \forall x_4 \ldots \exists x_n (\varphi(x_1, x_2, \ldots, x_n, x) = m)$$

defines a set of natural numbers with respect to the free variable x. Suitably modified, this idea will yield a powerful extension of the concept introduced in §3.

Chapter II

FINITE PROBABILITY FIELDS

In this chapter we present the elements of a constructive theory of probability, at this stage only for a finite space of elementary events. In view of this reservation, the definitions and results will be stated in rather less generality than would be possible in a discussion of constructive probability theory in general. Our construction largely follows Loève (1960) and Feller (1950); we have also been influenced by the intuitionistic ideas of Dijkman (1965).

§1. RANDOM EVENTS AND PROBABILITY

In the sequel we shall use capital italics A, B, C, E and H (with or without subscripts) to denote fixed finite sets of words over an alphabet B. Note that these letters are not designations of the predicates representing the sets, and it is meaningless to enclose them in curly brackets.

The symbol X will be used to denote a variable whose values are finite sets.

Definition 1. Let E be a nonempty finite set and $\mathfrak{B}$ an algorithm applicable to any element $e \in E$, producing a real number $\mathfrak{B}(e)$ such that

(1) $0 \not> \mathfrak{B}(e) \not> 1$;

(2) $\sum_{e \in E} \mathfrak{B}(e) = 1$.

The system $\Omega = (E, \mathfrak{B})$ will be called a probability field (PF), and we shall call E the space of elementary events of the field.

Definition 2. A random event in the field Ω is any finite subset of E (including E itself and the empty set $\varnothing$).

A random event will be called an elementary event if its norm is 1. For brevity's sake, we shall sometimes specify random events by simply listing their elements. Moreover, we shall assume that the elements of the set E are numbered from 1 to $\mu(E)$. The i-th element of E will be denoted by e_i. This will enable us, if necessary, to work not with the elements e themselves but rather with their indices i, $1 \leqslant i \leqslant \mu(E)$.

Two events* A and B are said to be equal (notation $A \approx B$) if they are equal as sets. The sum or union of events A and B is defined to be the set $C \approx A \cup B$. The product or intersection of events A and B is defined to be the set $D \approx AB$. The complement of an event A is the complement

* We shall now drop the word "random," since only random events will be considered in this chapter.

of the set A with respect to E, i. e., the set $\overline{A}E$. Using a slight abuse of notation, we shall henceforth denote the complement of an event A by $\overline{A}$. It is easy to see that the sum and product of two events are events, and the complement of an event is an event. By virtue of these definitions, all results concerning the classical algebra of events carry over without change to the constructive algebra of events, and we shall utilize such results in the sequel without proof.

Definition 3. The probability of an event A in the probability field Ω is defined as the number

$$p(A) = \sum_{e \in A} \mathfrak{B}(e);$$

if $A = \varnothing$, we set $p(A) = 0$.

The following properties are readily proved:

$$p(A \cup B) = p(A) + p(B) - p(AB);$$

$$p(\overline{A}) = 1 - p(A);$$

$$AB = \varnothing \supset p(A \cup B) = p(A) + p(B).$$

Definition 4. Let A, H be two events in the probability field Ω such that $p(H) > 0$. Then the (conditional) probability of A given H is defined as

$$p(A/H) = \frac{p(AH)}{p(H)}$$

The proof of the following theorem is almost the same as in the classical theory (see Feller, 1950).

Theorem 1. Let A and $H_1, \ldots, H_c$ be events such that

$$H_1 \cup \ldots \cup H_c = E,$$

and for $i \neq j$,

$$H_i H_j = \varnothing;$$

then

$$p(A) = \sum_{i=1}^{c} p(A/H_i)\, p(H_i).$$

§2. RANDOM VARIABLES AND DISTRIBUTIONS

Definition 5. Given a probability field $\Omega = (E, \mathfrak{B})$. A random variable over Ω, or simply a random variable, is defined to be an algorithm $\mathfrak{P}$, applicable to any element $e \in E$ and producing a real number.

The numbers $\mathfrak{P}(e)$ will be referred to as the values of the random variable $\mathfrak{P}$. The number $p(e)$ is called the probability of the value $\mathfrak{P}(e)$ at the point e. If all values of $\mathfrak{P}$ are rational, we can go even further and define the probability that $\mathfrak{P}$ will assume a value $a=\mathfrak{P}(e)$ in general, not only at a given point e. This is possible because the set $\{\mathfrak{P}(x)=b\}$, where b is a value of $\mathfrak{P}$, is finite.

The number $p(\{\mathfrak{P}(x)=a\})$ will be called the probability that the random variable $\mathfrak{P}$ assumes the value a.

Definition 6. The probability distribution, or simply distribution, of a random variable $\mathfrak{P}$ over Ω, given that the set $\{\mathfrak{P}(x)=\mathfrak{P}(e)\}$ is finite for any $e\in E$, is defined as the list of all distinct number pairs

$$(p(\{\mathfrak{P}(x)=\mathfrak{P}(e)\}), \mathfrak{P}(e)).$$

According to this definition, the distribution of a random variable $\mathfrak{P}$ over Ω is undefined unless one has an effective procedure $\Phi_{\mathfrak{P}}$ to decide which of the components of every disjunction

$$(\mathfrak{P}(e_i)=\mathfrak{P}(e_j))\vee(\mathfrak{P}(e_i)\neq\mathfrak{P}(e_j)),$$
$$e_i,\ e_j\in E,$$

is true. Since every disjunction of this type may be encoded as a well-defined word over a suitable alphabet B', the "effective procedure" $\Phi_{\mathfrak{P}}$ should, strictly speaking, be a normal algorithm over B'. For example, if the disjunctions are represented by words $e_i\delta e_j$ over the alphabet

$$B'=(\xi_1, \ldots, \xi_n, \delta),$$

the method $\Phi_{\mathfrak{P}}$ will be a normal algorithm over B', applicable to any word $e_i\delta e_j$ and producing the empty word if and only if the statement $\mathfrak{P}(e_i)=\mathfrak{P}(e_j)$ is true. This raises an interesting question: is there a normal algorithm applicable to any random variable $\mathfrak{P}$ over a given field Ω and producing such a "procedure"? This question is of course meaningful only when the alphabet is fixed for all random variables $\mathfrak{P}$ under consideration and their schemes are encoded as words over a suitable alphabet. It is easy to see that this problem is undecidable. Indeed, let $\Omega=(E, \mathfrak{B})$ be a probability field in which E contains exactly two elements. With every real number a, associate the random variable $\mathfrak{P}_a$ defined by

$$\begin{cases} \mathfrak{P}_a(e_1)=1; \\ \mathfrak{P}_a(e_2)=a. \end{cases} \qquad (\alpha)$$

Then the assumption that there exists an algorithm as described above leads to a contradiction with the true formula

$$\neg\forall a(a=1\vee a\neq 1).$$

§ 3. MEAN AND VARIANCE

Definition 7. Let $\mathfrak{P}$ be a random variable over a probability field $\Omega(E, \mathfrak{B})$; the number

$$M(\mathfrak{P}) = \sum_{e \in E} \mathfrak{P}(e)\mathfrak{B}(e)$$

will be called the mean of $\mathfrak{P}$ and

$$D(\mathfrak{P}) = \sum_{e \in E} (\mathfrak{P}(e) - M(\mathfrak{P}))^2 \mathfrak{B}(e)$$

its variance. Let $\mathfrak{P}$ be a random variable and a a real number. Let $a\mathfrak{P}$ and $\mathfrak{P}^2$ denote the random variables with values $a\mathfrak{P}(e)$ and $(\mathfrak{P}(e))^2$, respectively. The following properties are easily proved:

$$M(a\mathfrak{P}) = aM(\mathfrak{P});$$

$$D(a\mathfrak{P}) = a^2 D(\mathfrak{P});$$

$$D(\mathfrak{P}) = M(\mathfrak{P}^2) - (M(\mathfrak{P}))^2.$$

Let $\mathfrak{P}_1, \mathfrak{P}_2, \ldots, \mathfrak{P}_k$ be random variables over Ω. The sum $\mathfrak{P}_1 + \mathfrak{P}_2 + \ldots + \mathfrak{P}_k$ and product $\mathfrak{P}_1\mathfrak{P}_2 \ldots \mathfrak{P}_k$ of $\mathfrak{P}_1, \mathfrak{P}_2, \ldots, \mathfrak{P}_k$ are defined as random variables $\mathfrak{F}$ and , respectively, such that

$$\mathfrak{F}(e) = \sum_{i=1}^{k} \mathfrak{P}_i(e), \tag{1}$$

$$\mathfrak{F}'(e) = \prod_{i=1}^{k} \mathfrak{P}_i(e). \tag{2}$$

It follows from (1) and Definition 7 that

$$M(\mathfrak{P}_1 + \mathfrak{P}_2 + \ldots + \mathfrak{P}_k) = M(\mathfrak{P}_1) + M(\mathfrak{P}_2) + \ldots + M(\mathfrak{P}_k). \tag{3}$$

We may view any constant a, i. e., fixed real number, as a random variable all of whose values are equal to a; this convention will save space and simplify the formulas.

The difference of two random variables $\mathfrak{P}_1$ and $\mathfrak{P}_2$ over Ω is defined to be the random variable $\mathfrak{P}_1 + ((-1)\mathfrak{P}_2)$ (notation: $\mathfrak{P}_1 - \mathfrak{P}_2$).

Let $\mathfrak{P}$ be a random variable over $\Omega = (E, \mathfrak{B})$. We let $|\mathfrak{P}|$ denote the random variable with values $|\mathfrak{P}(e)|$.

Definition 8. Let $\mathfrak{P}_1$ and $\mathfrak{P}_2$ be random variables over Ω such that

$$M(\mathfrak{P}_1) = \mu_1;$$

$$M(\mathfrak{P}_2) = \mu_2;$$

the *covariance* of $\mathfrak{P}_1$, $\mathfrak{P}_2$ is defined to be the number

$$\operatorname{Cov}(\mathfrak{P}_1, \mathfrak{P}_2) = M((\mathfrak{P}_1 - \mu_1)(\mathfrak{P}_2 - \mu_2)).$$

Simple arguments show that

$$\operatorname{Cov}(\mathfrak{P}_1, \mathfrak{P}_2) = M(\mathfrak{P}_1\mathfrak{P}_2) - \mu_1\mu_2. \tag{4}$$

By (3), the following formula may be proved as in the classical case (see Feller, 1950):

$$D(\mathfrak{P}_1 + \mathfrak{P}_2 + \ldots + \mathfrak{P}_k) = D(\mathfrak{P}_1) + D(\mathfrak{P}_2) + \ldots + D(\mathfrak{P}_k) + \\ + 2\sum_{i,j} \operatorname{Cov}(\mathfrak{P}_i, \mathfrak{P}_{i+j}), \tag{5}$$

where the summation is over $i+j \leqslant k$.

Formula (5) may also be proved without appeal to (3), by removing parentheses and regrouping terms in the sum defining $D(\mathfrak{P}_1 + \mathfrak{P}_2 + \ldots + \mathfrak{P}_k)$.

§4. INDEPENDENT RANDOM VARIABLES

Let $\mathfrak{P}$ be a random variable over $\Omega = (E, \mathfrak{B})$ and a one of its values. We let $\{\mathfrak{P} = a\}$ denote the set of elementary events $E\{\mathfrak{P}(x) = a\}$. As we have seen (§2), the question whether $\{\mathfrak{P} = a\}$ may legitimately be considered an event must be settled separately in each individual case.

Now let $\mathfrak{P}_1, \mathfrak{P}_2, \ldots, \mathfrak{P}_k$ be random variables over Ω and $a_1, a_2, \ldots, a_k$ certain of their values. We introduce the notation $\{\mathfrak{P}_1 = a_1, \mathfrak{P}_2 = a_2, \ldots, \mathfrak{P}_k = a_k\}$ for the set of elementary events

$$\{\mathfrak{P}_1 = a_1\}\{\mathfrak{P}_2 = a_2\} \ldots \{\mathfrak{P}_k = a_k\}.$$

It is quite easy to disprove the following assertion: If $\{\mathfrak{P}_1 = a_1, \mathfrak{P}_2 = a_2, \ldots, \mathfrak{P}_k = a_k\}$ is an event, then so are

$$\{\mathfrak{P}_1 = a_1\}, \{\mathfrak{P}_2 = a_2\}, \ldots, \{\mathfrak{P}_k = a_k\}.$$

To do so, we consider the random variables $\mathfrak{P}_a$ over Ω as given by (α) in §2. Clearly, for any value of a the set $\{\mathfrak{P}_a = 1, \mathfrak{P}_0 = 1\}$ is an event, but the set $\{\mathfrak{P}_a = 1\}$ is in general not an event; to be precise, the statement that the set $\{\mathfrak{P}_a = 1\}$ is an event for any a turns out to be false. The situation is otherwise if we assume that all sets of type $\{\mathfrak{P}_1 = a_1, \mathfrak{P}_2 = a_2, \ldots, \mathfrak{P}_k = a_k\}$ are events. In other words, we can prove the following proposition:

Suppose that for any k-tuple $(a_1, a_2, \ldots, a_k)$ *of values of* $\mathfrak{P}_1, \mathfrak{P}_2, \ldots, \mathfrak{P}_k$ *the set* $\{\mathfrak{P}_1 = a_1, \mathfrak{P}_2 = a_2, \ldots, \mathfrak{P}_k = a_k\}$ *is an event; then the sets* $\{\mathfrak{P}_1 = a_1\}$, $\{\mathfrak{P}_2 = a_2\}, \ldots, \{\mathfrak{P}_k = a_k\}$ *are also events.*

Indeed, by assumption, for any value a_j of the random variable $\mathfrak{P}_j$ we have

$$\forall e(e \in \{\mathfrak{P}_1=\mathfrak{P}_1(e),\ \mathfrak{P}_2=\mathfrak{P}_2(e), \ldots, \mathfrak{P}_j=\mathfrak{P}_j(e), \ldots, \mathfrak{P}_k=\mathfrak{P}_k(e)\} \vee$$
$$\vee e \notin \{\mathfrak{P}_1=\mathfrak{P}_1(e),\ \mathfrak{P}_2=\mathfrak{P}_2(e), \ldots, \mathfrak{P}_j=\mathfrak{P}_j(e), \ldots, \mathfrak{P}_k=\mathfrak{P}_k(e)\}).$$

Hence, by the definition of these sets, we must have

$$\forall e\{\mathfrak{P}_j(e)=a_j \vee \mathfrak{P}_j(e) \neq a_j\}.$$

Consequently, for any $e \in E$ we can decide which component of the disjunction

$$e \in \{\mathfrak{P}_j=a_j\} \vee e \notin \{\mathfrak{P}_j=a_j\}$$

is true. This is equivalent to the statement that the set $\{\mathfrak{P}=a_j\}$ is finite, proving our assertion.

Definition 9. We shall say that events $A_1, A_2, \ldots, A_k$ in a field Ω are *independent* if, for any m-tuple $(m \leqslant k)$ of pairwise distinct natural numbers c_i, $1 \leqslant c_i \leqslant k$,

$$p(A_{c_1}A_{c_2}\ldots A_{c_m})=p(A_{c_1}) \cdot p(A_{c_2}) \ldots p(A_{c_m}).$$

The next definition refers to the situation in which all random variables in question have distributions, i. e., the sets $\{\mathfrak{P}=a\}$ are events.

Definition 10. Random variables $\mathfrak{P}_1, \mathfrak{P}_2, \ldots, \mathfrak{P}_k$ over Ω are said to be *independent* if, for any k-tuple of their respective values, the events $\{\mathfrak{P}_1=a_1\}, \{\mathfrak{P}_2=a_2\}, \ldots, \{\mathfrak{P}_k=a_k\}$ are independent.

It follows from the definition of independent random variables that for any k-tuple $(a_1, a_2, \ldots, a_k)$ of values of $\mathfrak{P}_1, \mathfrak{P}_2, \ldots, \mathfrak{P}_k$

$$p(\{\mathfrak{P}_1=a_1,\ \mathfrak{P}_2=a_2, \ldots, \mathfrak{P}_k=a_k\})=$$
$$=p(\{\mathfrak{P}_1=a_1\}) \cdot p(\{\mathfrak{P}_2=a_2\}) \ldots p(\{\mathfrak{P}_k=a_k\}).$$

Let $\mathfrak{P}_1$ and $\mathfrak{P}_2$ be independent random variables over Ω. Let b_i, $i=1, 2, \ldots, m$, and b'_j, $j=1, 2, \ldots, m'$, be all the distinct values of $\mathfrak{P}_1$ and $\mathfrak{P}_2$, respectively. Then

$$M(\mathfrak{P}_1)M(\mathfrak{P}_2)=\left(\sum_{e \in E}\mathfrak{P}_1(e)\mathfrak{B}(e)\right)\left(\sum_{e \in E}\mathfrak{P}_2(e)\mathfrak{B}(e)\right)=$$
$$=\left(\sum_{i=1}^{m} b_i p(\{\mathfrak{P}_1=b_i\})\right)\left(\sum_{j=1}^{m'} b'_j p(\{\mathfrak{P}_2=b'_j\})\right)=$$
$$=\sum_{i,j} b_i b'_j p(\{\mathfrak{P}_1=b_i,\ \mathfrak{P}_2=b'_j\})=\sum_{e \in E}\mathfrak{P}_1(e)\mathfrak{P}_2(e)\mathfrak{B}(e).$$

Consequently, the means of independent random variables $\mathfrak{P}_1$, $\mathfrak{P}_2$ satisfy the formula

$$M(\mathfrak{P}_1\mathfrak{P}_2)=M(\mathfrak{P}_1)M(\mathfrak{P}_2). \qquad (6)$$

Formulas (4), (5) and (6) show that the Bienaymé equality is valid for independent random variables $\mathfrak{P}_1, \mathfrak{P}_2, \ldots, \mathfrak{P}_k$ over Ω:

$$D(\mathfrak{P}_1+\mathfrak{P}_2+\ldots+\mathfrak{P}_k)=D(\mathfrak{P}_1)+D(\mathfrak{P}_2)+\ldots+D(\mathfrak{P}_k). \tag{7}$$

Indeed, to prove (7) we need only pairwise independence of the random variables.

§5. FIELD OF INDEPENDENT TRIALS

We now assume that $\Omega=(E, \mathfrak{B})$ is a probability field and ξ_δ a letter of the alphabet B which does not occur in any element $e\in E$.

Definition 11. A probability field $\Omega'=(E', \mathfrak{B}')$ is the *field of r-independent trials* associated with $\Omega=(E, \mathfrak{B})$ if E' is the set of all words e' of the form

$$\xi_\delta e_{i_1}\xi_\delta e_{i_2}\xi_\delta\ldots\xi_\delta e_{i_r}\xi_\delta,$$

where $e_{i_k}\in E$, $k=1,\ldots,r$, and $\mathfrak{B}'$ is defined by

$$\mathfrak{B}'(e')=\mathfrak{B}(e_{i_1})\cdot\mathfrak{B}(e_{i_2})\ldots\mathfrak{B}(e_{i_r}).$$

If E contains only two elements e_1, e_2, Ω' is called the field of Bernoulli r-trials.

Let Ω' be the field of Bernoulli r-trials associated with a field Ω. Let p and q denote $\mathfrak{B}(e_1)$ and $\mathfrak{B}(e_2)$, respectively, and $\mathfrak{P}^*(e')$ the number of occurrences of the word $\xi_\delta e_1\xi_\delta$ in e'. The function $\mathfrak{P}^*(e')$ is clearly a random variable $\mathfrak{P}^*$ over Ω', equal to the sum of r random variables $\mathfrak{P}_j$ defined as follows:

$$\mathfrak{P}_j(\xi_\delta e_{i_1}\xi_\delta e_{i_2}\xi_\delta\ldots\xi_\delta e_{i_r}\xi_\delta)=1$$

if $e_{i_j}\equiv e_1$; and otherwise

$$\mathfrak{P}_j(\xi_\delta e_{i_1}\xi_\delta e_{i_2}\xi_\delta\ldots\xi_\delta e_{i_n}\xi_\delta)=0.$$

Using familiar arguments, one can prove that

$$p(\{\mathfrak{P}^*=k\})=C_r^k p^k q^{r-k}. \tag{8}$$

It is easily seen that $\mathfrak{P}_j$, $j=1, 2, \ldots, r$, form an independent system of random variables, and for any j we have

$$p(\{\mathfrak{P}_j=1\})=p.$$

Consequently, for any j,

$$M(\mathfrak{P}_j)=p; \tag{9}$$

$$D(\mathfrak{P}_j)=pq. \tag{10}$$

22098

By (3), (7), (9) and (10), the mean and variance of the random variable $\mathfrak{P}^*$ are given by

$$M(\mathfrak{P}^*) = rp; \tag{11}$$

$$D(\mathfrak{P}^*) = rpq. \tag{12}$$

Using the properties of random variables $a\mathfrak{P}$ and formulas (11), (12), we obtain

$$M(r^{-1}\mathfrak{P}^*) = p; \tag{11'}$$

$$D(r^{-1}\mathfrak{P}^*) = \frac{pq}{r}. \tag{12'}$$

§6. CHEBYSHEV'S INEQUALITY

The celebrated Chebyshev inequality plays a major role in classical probability theory. It is therefore natural to ask whether an analogous theorem is valid in the constructive theory. It is however readily understood that a direct translation of the classical Chebyshev inequality into the constructive context is devoid of any meaning. Indeed, we would have to stipulate that for any positive number t and random variable $\mathfrak{P}$ over Ω the set $\{|\mathfrak{P} - M(\mathfrak{P})| > t\}$ is an event; this is inadmissible, for no general proof that such sets are finite is forthcoming.

Without a proof of this kind, we cannot legitimately refer to the probability that a random variable $\mathfrak{P}$ assumes a value satisfying the necessary conditions. In the constructive theory, then, Chebyshev's inequality must be given some other formulation. One of these is the following.

Let $\mathfrak{P}$ be a random variable over Ω and $t > 0$ a real number such that $\{|\mathfrak{P} - M(\mathfrak{P})| > t\}$ is an event. Then

$$p(\{|\mathfrak{P} - M(\mathfrak{P})| > t\}) \not> \frac{D(\mathfrak{P})}{t^2}.$$

The proof is essentially the same as in the classical case (see, e. g., Feller, 1950).

In practice, however, one is frequently in a position where it would be convenient to regard certain sets of elementary events, which are not events in the strict sense (see §1), as events nonetheless. "Officially," we have no right to speak of their probabilities until we establish that the corresponding sets are finite. Nevertheless, it is sometimes convenient to speak of the probability of some "event" without awaiting a logical justification of this action in the framework of our present conceptions of probability theory. In fact, the classical interpretation of Chebyshev's inequality enables one to estimate the probability of such "events" which certainly cannot be said of the constructive framework discussed above.

For these reasons, we have introduced a new system of concepts and notation, aimed at a more adequate analysis of the constructive content of Chebyshev's inequality.

Definition 12. If $\Omega=(E, \mathfrak{B})$ is a probability field, we define a quasi-event to be any set of elements $e \in E$ (including E itself and the empty set $\varnothing$).

Quasi-events will always be denoted by upper case Greek letters Γ, Δ (with or without subscripts).

We shall say that the probability of a quasi-event Γ is at most c if

$$\forall X (X \subseteq \overline{\overline{\Gamma}} \supset (p(X) \not> c)). \tag{13}$$

Similarly, the probability of a quasi-event Γ is at least c if

$$\forall X (X \subseteq \overline{\Gamma} \supset (p(X) \not> 1-c)). \tag{14}$$

As abbreviations for formulas (13) and (14) we shall write $p_\Gamma \dot{\not>} c$ and $p_\Gamma \dot{\not<} c$, respectively.

Finally, we shall say that the probability of a quasi-event is c if

$$(p_\Gamma \dot{\not>} c) \,\&\, (p_\Gamma \dot{\not<} c).$$

Our abbreviated notation for this situation will be $p_\Gamma \dot{=} c$. In our new system of notation, Chebyshev's inequality is formulated as follows: for any random variable $\mathfrak{P}$ over Ω and any real number $t>0$,

$$p_{\{|\mathfrak{P}-M(\mathfrak{P})|>t\}} \dot{\not>} \frac{D(\mathfrak{P})}{t^2}. \tag{15}$$

We first observe that

$$\overline{\overline{\{|\mathfrak{P}-M(\mathfrak{P})|>t\}}} \approx \{|\mathfrak{P}-M(\mathfrak{P})|>t\}. \tag{α'}$$

This is easily deduced from the following theorem of constructive analysis:

$$\forall \alpha \, \forall \beta ((\alpha > \beta) \leftrightarrow \neg\neg(\alpha > \beta)),$$

where α and β are variables for constructive real numbers.

Now let A be an event contained in $\{|\mathfrak{P}-M(\mathfrak{P})|>t\}$:

$$A \subseteq \{|\mathfrak{P}-M(\mathfrak{P})|>t\}.$$

If A is empty, we have

$$p(A)=0 \not> \frac{D(\mathfrak{P})}{t^2}. \tag{β'}$$

Now suppose that A is not empty; then

$$D(\mathfrak{P}) = \sum_{e \in E} (\mathfrak{P}(e) - M(\mathfrak{P}))^2 \mathfrak{B}(e) \nless$$

$$\nless \sum_{e \in A} (\mathfrak{P}(e) - M(\mathfrak{P}))^2 \mathfrak{B}(e) \nless t^2 \sum_{e \in A} \mathfrak{B}(e) = t^2 p(A).$$

Consequently,

$$p(A) \ngtr \frac{D(\mathfrak{P})}{t^2}.$$

Together with (α'), this gives

$$\forall X (X \subseteq \{|\mathfrak{P} - M(\mathfrak{P})| > t\} \supset p(X) \ngtr \frac{D(\mathfrak{P})}{t^2}. \qquad (\gamma')$$

Formula (15) now follows directly from (α'), (γ') and (13).

Using (11'), (12') and (15), we can prove a formula which may be interpreted as the law of large numbers:

$$p_{\{|r^{-1}\mathfrak{P}^* - p| > t\}} \dot{\ngtr} \frac{pq}{t^2 r}. \qquad (16)$$

Letting $\varepsilon = \frac{1}{2} - p$, one easily proves that

$$pq = \left(\frac{1}{2} - \varepsilon\right)\left(\frac{1}{2} + \varepsilon\right) \ngtr \frac{1}{4}.$$

Thus the expression on the right of (16) may be made independent of p, replacing pq by its maximum value $\frac{1}{4}$. Consequently, for any $t > 0$,

$$p_{\{|r^{-1}\mathfrak{P}^* - p| > t\}} \dot{\ngtr} \frac{1}{4t^2 r}. \qquad (16')$$

§7. KOLMOGOROV'S INEQUALITY

Kolmogorov's theorem. Let $\mathfrak{P}_1, \mathfrak{P}_2, \ldots, \mathfrak{P}_n$ *be independent random variables over*

$$\Omega = (E, \mathfrak{B});$$

$$\mathfrak{S}_i = \mathfrak{P}_1 + \ldots + \mathfrak{P}_i;$$

$$M(\mathfrak{S}_i) = \mu_i, \qquad i = 1, 2, \ldots, n,$$

and

$$D(\mathfrak{S}_i) = s_i^2, \qquad i = 1, 2, \ldots, n.$$

Let $t>0$ be a real number and suppose that $s_n>0$. Then

$$p\{(|\mathfrak{S}_1-\mu_1|<ts_n)\&(|\mathfrak{S}_2-\mu_2|<ts_n)\&\ldots\&(|\mathfrak{S}_n-\mu_n|<ts_n)\}\not<1-t^{-2}. \tag{17}$$

Proof. The classical proof (see Feller, 1950) does not carry over directly, since the random variables Y_ν defined by

$$Y_\nu(e)=\begin{cases}1, & \text{if} \quad (\overset{\nu-1}{\underset{i=1}{\&}}(|\mathfrak{S}_i(e)-\mu_i|<ts_n))\&(|\mathfrak{S}_\nu(e)-\mu_\nu|\not<ts_n);\\ 0 & \text{otherwise.}\end{cases}$$

are actually not random variables in the constructive sense — we have no general algorithm computing the values $Y_\nu(e)$. Nonetheless, the main idea of the classical proof does carry over, as is readily perceived from the following constructive proof of (17).

For any natural number m we shall define a random variable $\mathfrak{R}_i^m$ in terms of the random variable $|\mathfrak{S}_i-\mu_i|-ts_n$ over Ω. To do this, we assume that the number

$$a=|\mathfrak{S}_i(e)-\mu_i|-ts_n, \quad e\in E,$$

is the code of a pair of algorithms $(\mathfrak{A}_e, \mathfrak{B}_e)$. By virtue of our conventions (see Chap. I), the algorithm $\mathfrak{A}_e$ determines a sequence of rational numbers, while $\mathfrak{B}_e$ is a convergence regulator for this sequence. We now define $\mathfrak{R}_i^m$ by

$$\mathfrak{R}_i^m(e)=\mathfrak{A}_e(\mathfrak{B}_e(m))+2^{-m}.$$

It is readily shown that the values of the random variables $\mathfrak{R}_i^m$ are rational numbers. This justifies our defining new random variables $\mathfrak{D}_\nu^m$ over Ω by

$$\mathfrak{D}_\nu^m(e)=\begin{cases}1, & \text{if} \quad \left(\overset{\nu-1}{\underset{i=1}{\&}}\mathfrak{R}_i^m(e)<0\right)\&\mathfrak{R}_\nu^m(e)\not<0;\\ 0 & \text{otherwise.}\end{cases}$$

Clearly, for any fixed m, we have the implication

$$e\in\overline{\left\{\overset{n}{\underset{i=1}{\&}}(|\mathfrak{S}_i-\mu_i|<ts_n)\right\}}\supset\exists\,\nu(\mathfrak{D}_\nu^m(e)=1), \quad 1\leqslant\nu\leqslant n. \tag{18}$$

It is also easy to see that for no pair of values (i, j), $1\leqslant i<j\leqslant n$, can the statement

$$\mathfrak{D}_i^m(e)=\mathfrak{D}_j^m(e)=1$$

be true. Hence it follows that the random variable

$$\mathfrak{D}^m=\mathfrak{D}_1^m+\mathfrak{D}_2^m+\ldots+\mathfrak{D}_n^m$$

takes only two values, 0 and 1. Now,

$$M((\mathfrak{S}_n-\mu_n)^2\mathfrak{D}^m)\not< M((\mathfrak{S}_n-\mu_n)^2\cdot 1)=s_n^2.$$

Define random variables $\mathfrak{A}_k$, $k=1, 2, \ldots, n-1$, by

$$\mathfrak{A}_k=(\mathfrak{S}_n-\mu_n)-(\mathfrak{S}_k-\mu_k)=\sum_{i=k+1}^{n}(\mathfrak{P}_i-M(\mathfrak{P}_i)).$$

It can be shown that

$$M((\mathfrak{S}_n-\mu_n)^2\mathfrak{D}_k{}^m)\not> M((\mathfrak{S}_k-\mu_k)^2\mathfrak{D}_k{}^m). \tag{19}$$

Indeed,

$$\begin{aligned}M((\mathfrak{S}_n-\mu_n)^2\mathfrak{D}_k{}^m)&=M((\mathfrak{S}_k-\mu_k+\mathfrak{A}_k)^2\mathfrak{D}_k{}^m)=\\&=M((\mathfrak{S}_k-\mu_k)^2\mathfrak{D}_k{}^m)+2M((\mathfrak{S}_k-\mu_k)\mathfrak{A}_k\mathfrak{D}_k{}^m)+\\&\quad+M((\mathfrak{A}_k)^2\mathfrak{D}_k{}^m).\end{aligned} \tag{20}$$

It is obvious that $\mathfrak{D}_k{}^m$ depends only on the first k random variables $\mathfrak{P}_i$. The same holds for the random variable $(\mathfrak{S}_k-\mu_k)$. But $\mathfrak{A}_k$ depends only on the last $n-k$ random variables $\mathfrak{P}_i$. Since the variables $\mathfrak{P}_i$ are independent, so are the random variables $\mathfrak{A}_k$ and $(\mathfrak{S}_k-\mu_k)\mathfrak{D}_k{}^m$. Consequently,

$$\begin{aligned}&M((\mathfrak{S}_k-\mu_k)\mathfrak{A}_k\mathfrak{D}_k{}^m)=\\&=M((\mathfrak{S}_k-\mu_k)\mathfrak{D}_k{}^m)M(\mathfrak{A}_k)=\\&=M((\mathfrak{S}_k-\mu_k)\mathfrak{D}_k{}^m)\cdot 0=0.\end{aligned} \tag{21}$$

Formula (19) now follows at once from (20), (21) and the fact that the values of the random variable $(\mathfrak{A}_k)^2\mathfrak{D}_k{}^m$ are negative.

It follows from the definition of $\mathfrak{R}_k{}^m$ that

$$|\mathfrak{S}_k(e)-\mu_k|-ts_n\not>\mathfrak{R}_k{}^m(e)-2^{-m+1}.$$

But if $\mathfrak{D}_k{}^m(e)=1$, then $\mathfrak{R}_k{}^m(e)\geqslant 0$. Consequently,

$$\forall e((\mathfrak{D}_k{}^m(e)=1)\supset|\mathfrak{S}_k(e)-\mu_k|\not> ts_n-2^{-m+1}). \tag{22}$$

Combining (19) and (22), we obtain

$$\begin{aligned}s_n^2&\not> M(\mathfrak{D}^m(\mathfrak{S}_n-\mu_n)^2)\not>\sum_{k=1}^{n}M(\mathfrak{D}_k{}^m(\mathfrak{S}_k-\mu_k)^2)\not>\\&\not>\sum_{k=1}^{n}M(\mathfrak{D}_k{}^m(ts_n-2^{-m+1})^2)=(ts_n-2^{-m+1})^2p(\{\mathfrak{D}^m=1\})\end{aligned}$$

Hence, if $s_n\neq 0$, it follows that

$$p(\{\mathfrak{D}^m=1\})\not<\frac{1}{(t-2^{-m+1}s_n{}^{-1})^2}. \tag{23}$$

Supposing A to be an event such that

$$A \subseteq \{\overline{\mathop{\&}_{i=1}^{n} (|\mathfrak{S}_i - \mu_i| < ts_n)}\},$$

we see (see (18) and (23)) that for any m

$$p(A) \not> \frac{1}{(t - 2^{-m+1} s_n^{-1})^2}.$$

Consequently,

$$p(A) \not> \frac{1}{t^2},$$

which proves (17).

NOTES

The constructive theory of probability presented above is in many respects similar in approach to Dijkman's intuitionistic probability theory (Dijkman, 1965). Underlying this similarity is of course the relationship between constructive and intuitionistic mathematics. In spite of this, the system of constructive mathematics we have used gives rise to certain differences between these conceptions. We shall not dwell on these here; we shall confine ourselves only to some distinctive features that stem from another source.

Dijkman (1965) considers discrete state spaces which need not be finite sets. In this respect his intuitionistic theory generalizes our theory as presented above. Moreover, his concept of random event is different. Were we to use Dijkman's ideas here, the result would be the following definition:

Definition 13. A set $\{P\}$ is a *random event* in a probability field $\Omega = (E, \mathfrak{B})$ if and only if $\{P\} \subseteq E$ and

$$\forall_{e \in E} e(\mathfrak{B}(e) \neq 0 \supset (e \in \{P\} \vee e \in \{\overline{P}\})).$$

Naturally, any event in the sense of Definition 2 is an event in the sense of Definition 13, but the converse is false. For example, the set $\{\mathfrak{B} > 0\}$ in Ω is an event by Definition 13, but there is no general proof that any such set is finite. It can be shown nonetheless that the events in the sense of Definition 13 are quasi-events Γ with the property that there exists a number b with

$$p_\Gamma \doteq b. \tag{24}$$

Indeed, suppose that the number $\mathfrak{B}(e_i)$ is represented by a pair of algorithms $\mathfrak{A}_i$, $\mathfrak{B}_i$; for each natural number m, we define a random variable $\mathfrak{P}_m$ over Ω:

$$\mathfrak{P}_m(e_i) = \mathfrak{A}_i(\mathfrak{B}_i(m)) - 2^{-m}. \qquad (25)$$

The set $\{\mathfrak{P}_m > 0\}$ is clearly finite. It is readily seen that the sets $\overline{\overline{\Gamma}}\{\mathfrak{P}_m > 0\}$ and $\overline{\Gamma}\{\mathfrak{P}_m > 0\}$ are also finite, provided Γ is an event in the sense of Definition 13.

Construct sequences of numbers $\{c_m\}$ and $\{d_m\}$ by

$$c_m = p(\overline{\overline{\Gamma}}\{\mathfrak{P}_m > 0\});$$
$$d_m = p(\overline{\Gamma}\{\mathfrak{P}_m > 0\}).$$

By (24), the sequences $\{c_m\}$ and $\{d_m\}$ have limits, say

$$\lim_{m\to\infty} c_m = c,$$
$$\lim_{m\to\infty} d_m = d.$$

Then, in view of the relation

$$\{\mathfrak{P}_m > 0\} \approx \overline{\overline{\Gamma}}\{\mathfrak{P}_m > 0\} \cup \overline{\Gamma}\{\mathfrak{P}_m > 0\},$$

we have $d = 1 - c$. Now let A and B be two events such that $A \subseteq \overline{\overline{\Gamma}}$ and $B \subseteq \overline{\Gamma}$. It is easy to prove that

$$\lim_{m\to\infty} p(A\{\mathfrak{P}_m > 0\}) = p(A)$$

and

$$\lim_{m\to\infty} p(B\{\mathfrak{P}_m > 0\}) = p(B).$$

But since

$$A\{\mathfrak{P}_m > 0\} \subseteq \overline{\overline{\Gamma}}\{\mathfrak{P}_m > 0\}$$

and

$$B\{\mathfrak{P}_m > 0\} \subseteq \overline{\Gamma}\{\mathfrak{P}_m > 0\},$$

it follows that

$$p(A) \not> c$$

and

$$p(B) \not> 1 - c.$$

Thus the limit of the sequence $\{c_m\}$ satisfies (24).

On the other hand, there exists a class of quasi-events such that each of its members Γ satisfies the relation $p_\Gamma \doteq 1$, but it is essentially impossible to prove that every quasi-event of the class is an event in the sense of Definition 13. An example is the class of all events

$$\{(\mathfrak{P}=a) \vee (\mathfrak{P} \neq a)\},$$

where $\mathfrak{P}$ is a random variable over Ω and a an arbitrary real number.

These results seem to indicate that (24) should be used as the definition of an event. This approach, however, involves certain difficulties. For example, it is doubtful whether one could prove that the sum of any two events is again an event. In fact, this result is more likely to be false.

Differences arise in the definition of a random variable as well. Disregarding the fact that Dijkman considers a broader class of state spaces, we can assert that his definition is conceptually narrower than ours, for our definition of a random variable $\mathfrak{P}$ over Ω (see §2, Definition 5) involves no restrictive condition of the type

$$\forall i \, \forall j((\mathfrak{P}(e_i) = \mathfrak{P}(e_j)) \vee (\mathfrak{P}(e_i) \neq \mathfrak{P}(e_j))),^* \quad e_i, e_j \in E.$$

The concept of a weak stochastic variable, however, is in a certain sense more general than our concept of random variable.

Finally, the two approaches diverge as to their interpretation of Chebyshev's inequality (and, in a sense, of Kolmogorov's theorem). Dijkman's modification is a special case of our version.

* Using the method of constructive choice, one can prove the formula

$$(\mathfrak{P}(e_i) \neq \mathfrak{P}(e_j)) \supset ((\mathfrak{P}(e_i) < \mathfrak{P}(e_j)) \vee (\mathfrak{P}(e_i) > \mathfrak{P}(e_j))).$$

Chapter III

SIMPLE AND HOMOGENEOUS MARKOV CHAINS

In this chapter we shall use probability fields Ω having finite spaces of elementary events. Some important results in the theory of Markov chains will be proved only in later chapters. The reason for this is that certain results in the theory are special cases of theorems proved in the theory of FSA.

§1. STOCHASTIC MATRICES

A matrix

$$\mathbf{M}=\begin{pmatrix} a_{11} & a_{12} & \dots & a_{1n} \\ a_{21} & a_{22} & \dots & a_{2n} \\ \dots & \dots & \dots & \dots \\ \dots & \dots & \dots & \dots \\ a_{m1} & a_{m2} & \dots & a_{mn} \end{pmatrix},$$

whose entries are CRNs, will be called a *stochastic* $m\times n$ *matrix* if $a_{ij}\nless 0$ and $\forall\, i \left(\sum_{i=1}^{n} a_{ij}=1\right)$.

If $m=1$, we shall call $\mathbf{M}$ a stochastic row-vector, or simply a *stochastic vector*.

Property 1. Let $\mathbf{M}_1=(a_{ij})$ be a stochastic $m\times n$ matrix and $\mathbf{M}_2=(b_{jk})$ a stochastic $n\times m'$ matrix. Then the product $\mathbf{M}_1\mathbf{M}_2$ is a stochastic $m\times m'$ matrix.

Indeed, the matrix $\mathbf{M}\rightleftharpoons\mathbf{M}_1\mathbf{M}_2$ has as many rows (columns) as $\mathbf{M}_1$ ($\mathbf{M}_2$) has rows (columns). That $\mathbf{M}=(c_{ij})$ is a stochastic matrix follows from the relations

$$c_{ik}=\sum_{j=1}^{n} a_{ij}b_{jk};$$

$$\sum_{k=1}^{m} c_{ik}=a_{i1}\sum_{k=1}^{m} b_{1k}+\dots+a_{in}\sum_{k=1}^{m'} b_{nk}$$

Corollary. If $\mathbf{M}_1$ is a stochastic vector (i. e., $m=1$), the product $\mathbf{M}_1\mathbf{M}_2$ is also a stochastic vector (with m' components).

A stochastic $m\times n$ matrix $\mathbf{M}=(a_{ij})$ is said to be doubly stochastic if

$$\forall j\left(\sum_{i=1}^{m} a_{ij}=1\right).$$

Property 2. A matrix $\mathbf{M}=(a_{ij})$ is doubly stochastic only if it has the same number of rows as columns.

Indeed,

$$a_{i1}=1-\sum_{j=1}^{m-1} a_{ij}$$

for all $i=1, 2, \ldots, m$. Thus

$$\sum_{i=1}^{m} a_{i1}=\sum_{i=1}^{m}\left(1-\sum_{j=1}^{n-1} a_{ij}\right).$$

But since $\mathbf{M}$ is doubly stochastic, this implies

$$1=m-\sum_{j=1}^{n-1}\sum_{i=1}^{m} a_{ij}=m-(n-1).$$

Thus we must have $m-n=0$, proving the assertion.

Property 3. If $\mathbf{M}_1$ and $\mathbf{M}_2$ are doubly stochastic $n\times n$ matrices, their product is also doubly stochastic.

The proof is almost the same as that of Property 1.

A matrix $\mathbf{M}$ (stochastic or otherwise) is said to be positive if all its entries are positive CRNs. We shall say that $\mathbf{M}$ contains a positive column if it has at least one column of positive CRNs. The product of two positive stochastic $n\times n$ matrices is clearly a positive stochastic matrix. It is also easy to see that the product of two stochastic $n\times n$ matrices, each containing a positive column, is again a stochastic $n\times n$ matrix containing a positive column.

A stochastic $n\times n$ matrix $\mathbf{M}$ is said to be regular if there is a natural number k such that $\mathbf{M}^k$ contains a positive column.

Property 4. There is no algorithm which decides whether or not a stochastic matrix is regular [i.e., the set of regular stochastic matrices is nonrecursive].

Proof. Suppose we are given a normal algorithm $\mathfrak{B}$ over an alphabet $A_0=\{\xi_1, \xi_2\}$.

With each word g over A_0 we associate a real number a_g, as follows. If

$$\mathfrak{B}(g)\vdash g_1,\quad \mathfrak{B}(g_1)\vdash g_2, \ldots, \mathfrak{B}(g_{n-1})\vdash g_n,^{*}$$

* The symbolic notation $\mathfrak{B}(g)\vdash g'$ means that the normal algorithm effects simple translation (i.e., in one step, by application of a nonresultant substitution formula) of the word g into the word g'; $\mathfrak{B}(g)\vdash\cdot g'$ denotes a resultant translation.

where $g_1, \ldots, g_n$ are words over A_0, then the first $n+1$ terms of the constructive sequence of rational numbers defining a_g are $1, \frac{1}{2}, \ldots, \frac{1}{n+1}$; if $\mathfrak{B}(g_n) \models \cdot g_{n+1}$, or if no substitution formula of the normal algorithm is now applicable to g_n, then the $(n+k)$-th term $(k \geqslant 1)$ of this sequence is $\frac{1}{n+1}$. In case no substitution formula of the algorithm is applicable to the word g, the constructive sequence of rational numbers defining the number a_g is $1, 1, \ldots$

Any sequence of rational numbers $\{r_n\}$ defined in this way has the following property: for every pair of natural numbers n_1, n_2, $n_1 \leqslant n_2$, we have

$$|r_{n_1} - r_{n_2}| \leqslant \frac{1}{n_1}.$$

It follows that the algorithm defining this sequence of rationals relative to a fixed word g, combined with the identity algorithm $\mathfrak{Z}(n) = n$, defines the real number a_g. We have $a_g > 0$ if and only if the algorithm $\mathfrak{B}$ is applicable to the word g. Markov showed that one can construct a normal algorithm over the alphabet A_0 such that no normal algorithm $\mathfrak{A}_0$ over A_0, applicable to any word g over A_0, can decide whether or not $\mathfrak{B}_0$ is applicable to words g over A_0.

Suppose now that $\mathfrak{B}$ is the algorithm $\mathfrak{B}_0$, and denote the set of all CRNs a_g by $\{B_g\}$. There is clearly no algorithm that decides whether or not numbers in $\{B_g\}$ are equal to zero.

Since all numbers d, $d \in \{B_g\}$, lie in the closed interval $[0, 1]$, the matrix

$$\mathbf{C} = \begin{pmatrix} 1 & 0 \\ d & 1-d \end{pmatrix}$$

is stochastic. It is clearly regular if and only if $d > 0$. Indeed, if $d > 0$ the first column of $\mathbf{C}$ is positive and so the matrix is regular. If $d = 0$, then $1 - d = 1$ and $\mathbf{C}$ is the unit matrix, [which is of course not regular]. Thus the regularity of $\mathbf{C}$ implies $d \neq 0$. Hence the assumption that the regularity of $\mathbf{C}$ may be decided by some algorithm leads to a contradiction, since it is equivalent to the existence of an algorithm telling which of the numbers d vanish.

Any $n \times n$ matrix will also be called an n-th order matrix. If each row of a stochastic $n \times n$ matrix contains an entry 1, we shall call it a matrix of type DS.

Definition 1. Given a stochastic vector $\pi_0 = (p_1, \ldots, p_k)$ and an ordered string of stochastic $n \times n$ matrices $\Pi_0 = (\mathbf{P}_1, \ldots, \mathbf{P}_k)$, all of whose terms are of type DS. The pair $[\pi_0, \Pi_0]$ will be called a decomposition of a stochastic $n \times n$ matrix $\mathbf{P}$ if $\mathbf{P} = p_1\mathbf{P}_1 + \ldots + p_k\mathbf{P}_k$; the number k will be called the decomposition length.

The following proposition is due to Podnieks.

Property 5. Every stochastic $n \times n$ matrix $\mathbf{P}$ has a decomposition $[\pi_0, \Pi_0]$.

Proof. Let $\Pi_0=(\mathbf{P}_1, \ldots, \mathbf{P}_N)$, $N=n^n$, contain all stochastic $n\times n$ matrices $\mathbf{P}_k=(b_{ij}^{(k)})$ of type DS. We introduce the notation $\mathbf{M}_k=(a_{ij}^{(k)})$, $k=1, 2, \ldots, N+1$, for the matrices defined as follows:

$$\mathbf{M}_1=\mathbf{P}_1;$$

$$\mathbf{M}_{k+1}=\mathbf{M}_k-\alpha_k\mathbf{P}_k,$$

where

$$\alpha_k=\min\{a_{1j_1}^{(k)}, \ldots, a_{nj_n}^{(k)}\}$$

if

$$\min\{b_{1j_1}^{(k)}, \ldots, b_{nj_n}^{(k)}\}=1.$$

It is easy to see that

$$\min\{a_{1j_1}^{(k+1)}, \ldots, a_{nj_n}^{(k+1)}\}=0 \tag{1}$$

and

$$\mathbf{M}_{N+1}=\mathbf{P}-\alpha_1\mathbf{P}_1-\ldots-\alpha_N\mathbf{P}_N. \tag{2}$$

We claim that $\mathbf{M}_{N+1}=(0)$. Indeed, suppose that $a_{st}^{(N+1)}>0$ for some s, t; then

$$\forall i\left(\sum_{j=1}^{n} a_{ij}^{(N+1)}=d>0\right).$$

But then the i-th row $(i=1, \ldots, n)$ must contain an entry $a_{ik_i}^{(N+1)}>0$, and so

$$\min\{a_{1k_1}^{(N+1)}, \ldots, a_{nk_n}^{(N+1)}\}>0. \tag{3}$$

On the other hand, by (1), we have

$$\min\{a_{1j_1}^{(N+1)}, \ldots, a_{nj_n}^{(N+1)}\}=0 \tag{4}$$

for all sequences $(j_1\ldots j_n)$. This contradiction shows that always $a_{ij}^{(N+1)}\not>0$. But since $a_{ij}^{(N+1)}\not<0$, it follows that

$$a_{ij}^{(N+1)}=0. \tag{5}$$

This implies that $\pi_0=(\alpha_1, \alpha_2, \ldots, \alpha_N)$ and Π_0 form a decomposition of $\mathbf{P}$.

It is readily seen that this algorithm for decomposition of a stochastic $n\times n$ matrix, applied to stochastic matrices with rational elements, yields a decomposition of length $n(n-1)+1$. Metra (1970) showed that for every n one can construct a stochastic $n\times n$ matrix whose minimal decomposition length is $n(n-1)+1$. This matrix has the form $n^{-n}\mathbf{M}$, where

$$\mathbf{M}=\begin{pmatrix} 1 & 1 & \dots & 1 & n^n-(n-1)\cdot 1 \\ n & n & \dots & n & n^n-(n-1)n \\ n^2 & n^2 & \dots & n^2 & n^n-(n-1)n^2 \\ \cdot & \cdot & \cdot & \cdot & \cdot \\ n^{n-1} & n^{n-1} & \dots & n^{n-1} & n^n-(n-1)n^{n-1} \end{pmatrix}.$$

Podnieks has also provided a method whereby, given a number ε and a stochastic $m\times n$ matrix $\mathbf{P}$, one can construct a matrix $\mathbf{P}'$ whose entries differ from those of $\mathbf{P}$ by at most ε and which has the minimal decomposition length $m(n-1)+1$. Considering the matrices $\mathbf{M}=\begin{pmatrix} 1 & 0 \\ d & 1-d \end{pmatrix}$, $d\in\{B_g\}$, one readily shows that there is no algorithm that produces a decomposition of minimal length.

§2. FINITE SIMPLE AND HOMOGENEOUS MARKOV CHAINS

Definition 2. A system $[E, \pi_0, \mathbf{P}]$, where E is a finite ordered set* over an alphabet B, $\mu(E)=n$, $\pi_0=(p_1, \dots, p_n)$ a stochastic vector, and $\mathbf{P}$ an n-th order stochastic matrix, will be called a simple homogeneous Markov chain with state set E, initial distribution π_0, and transition matrix $\mathbf{P}$.

Definition 3. A probability field $\Omega=(E^r, \mathfrak{B})$ will be called a field of first Markov r-trials induced by a Markov chain $[E, \pi_0, \mathbf{P}]$ if Ω possesses the following properties:

(1) The elementary events of Ω are all possible words $\xi_\delta e_{i_1}\xi_\delta e_{i_2}\xi_\delta\dots\xi_\delta e_{i_r}\xi_\delta$, where ξ_δ is a letter not occurring in any e, $e\in E$; e_{i_j}, $j=1, \dots, r$, are elements of E.

(2) $\mathfrak{B}(\xi_\delta e_{i_1}\xi_\delta e_{i_2}\xi_\delta\dots\xi_\delta e_{i_r}\xi_\delta)=p_{i_1}p_{i_1i_2}\dots p_{i_{r-1}i_r}$, where p_{ij} is the (i, j)-th element of the matrix $\mathbf{P}$.**

It is evident that in a certain sense Ω generalizes the concept of the field of r-independent trials associated with a PF. This assertion is supported by the following arguments. Let $\mathfrak{P}_{ij}$ denote a normal algorithm applicable to any word $\xi_\delta e_{i_1}\xi_\delta e_{i_2}\xi_\delta\dots\xi_\delta e_{i_r}\xi_\delta$, producing the natural number 1 if $e_{i_j}\equiv e_i$, $1\leqslant i\leqslant n$, $1\leqslant j\leqslant r$, and 0 otherwise.

The algorithm $\mathfrak{P}_{ij}$ is clearly a random variable over Ω, with the property that $\{\mathfrak{P}_{ij}=1\}$ is an event. For brevity, we denote this set by A_{ij} and its probability by $p_i(j)$.

Simple arguments show the truth of the following relations:

$$\forall j\left(\bigcup_{i=1}^{n} A_{ij}=E^r\right);$$

$$\forall j\left(i\neq i'\supset A_{ij}A_{i'j}=\varnothing\right).$$

This implies that $\sum_{i=1}^{n} p_i(j)=1$.

* A finite ordered set is a finite set whose elements are linearly ordered.

** The subscripts i_j assigned to the letters e agree with their order in the set E.

Suppose now that the field Ω is such that

$$p_i(j) > 0, \quad i=1, \ldots, n; \quad j=1, \ldots, r.$$

Then we have a system of equalities

$$\begin{cases} p_i(1) = p_i; \\ p(A_{ij+1}/A_{kj}) = p_{ki}, \quad j<r. \end{cases} \tag{6}$$

We see that the connection between Ω and $[E, \pi_0, \mathbf{P}]$ is in a certain sense invertible; to be precise: one can first determine Ω, using only condition (1) for this purpose, and then check for the existence of π_0 and $\mathbf{P}$ such that (6) holds. If we find that probability fields $\Omega=(E^r, \mathfrak{B})$ and $\Omega'=(E^r, \mathfrak{B}')$ satisfy (6) for the same pair $(\pi_0, \mathbf{P})$, we are justified in calling Ω' a PF inducing Ω.

The following properties of $\Omega=(E^r, \mathfrak{B})$ are proved along the same lines as the analogous classical results:

$$p(A_{i_1 1} A_{i_2 2} \ldots A_{i_{r-1} r-1} A_{i_r r}) = p_{i_1} \cdot p_{i_1 i_2} \ldots p_{i_{r-1} i_r}; \tag{7}$$

$$(p_1(j), p_2(j), \ldots, p_n(j)) = \pi_0 \mathbf{P}^{j-1}; \tag{8}$$

$$p(A_{k\, j+l}/A_{ij}) = p_{ik}(l), \tag{9}$$

where $j+l \leqslant r$ and $p_{ik}(l)$ is the (i, k)-th element of the matrix $\mathbf{P}^l$.

Using the random variables $\mathfrak{P}_{ij}$ over $\Omega=(E^r, \mathfrak{B})$, we define a new random variable $\mathfrak{P}^r$ over Ω.

Let $d_1, d_2, \ldots, d_n$ be given CRNs. Let $\mathfrak{P}_j$ denote the random variable over Ω defined by

$$\mathfrak{P}_j = d_1 \mathfrak{P}_{1j} + d_2 \mathfrak{P}_{2j} + \ldots + d_n \mathfrak{P}_{nj}. \tag{10}$$

Then the random variable $\mathfrak{P}^r$ over Ω is defined by

$$\mathfrak{P}^r = \mathfrak{P}_1 + \mathfrak{P}_2 + \ldots + \mathfrak{P}_r. \tag{11}$$

It is clear that if $\mathfrak{P}^r$ has a distribution the same is true of all the $\mathfrak{P}_j$'s, and it is thus legitimate to speak of the probability of the events $\{\mathfrak{P}^r=a\}$ and $\{\mathfrak{P}_j=d\}$, where a and d are values of $\mathfrak{P}^r$ and $\mathfrak{P}_j$, respectively. We shall nevertheless avoid the assumption that random variables of the type $\mathfrak{P}^r$ have distributions.

For convenience in notation, we let $M(\mathfrak{P}^1)$ denote the sum $p_1(1)d_1 + \ldots$ $\ldots + p_n(1)d_n$.

Definition 1. Let $\mathfrak{P}^r$ be random variables defined for the same n-tuple $d_1, d_2, \ldots, d_n$. Then, if the sequence of means

$$M(\mathfrak{P}^1), M(\mathfrak{P}^2), \ldots, M(\mathfrak{P}^r), \ldots$$

converges to some limit $M(\mathfrak{P})$, we shall say that the Markov chain induced by Ω has a *final mean* for random variables of type $\mathfrak{P}^r$.

The definition of the final variance $\mathfrak{D}(\mathfrak{P})$ is analogous. Both final mean and final variance are limits of suitable sequences. The random variables of type $\mathfrak{P}^r$ and their means and variances (including final mean and variance) are convenient in characterizing Markov chains from the standpoint of unlimited evolution. For example, the question of how long a Markov chain remains in a state e_i is solved quite easily with the help of our random variables $\mathfrak{P}^r$.

Property 6. Let $\Omega = (E^r, \mathfrak{B})$ *be the field induced by* $[E, \pi_0, \mathbf{P}]$. *Then the mean of a random variable of type* $\mathfrak{P}^r$ *is*

$$M(\mathfrak{P}^r) = r(\pi_0 \mathbf{P}_r \mathbf{D}), \tag{12}$$

where $\mathbf{P}_r = (\mathbf{E} + \mathbf{P} + \ldots + \mathbf{P}^{r-1}) r^{-1}$, $\mathbf{E}$ *is the unit matrix of order* n, *and*

$$\mathbf{D} = \begin{pmatrix} d_1 \\ d_2 \\ \cdot \\ \cdot \\ \cdot \\ d_n \end{pmatrix}.$$

Proof. If $r=1$, formula (12) is valid by the definition of $M(\mathfrak{P}^1)$. Assuming that it is valid for r, we prove it for $r+1$. Indeed,

$$\begin{aligned} M(\mathfrak{P}^{r+1}) &= \sum_{i_1 i_2 \ldots i_{r+1}} (d_{i_1} + d_{i_2} + \ldots + d_{i_{r+1}}) p_{i_1}(1) p_{i_1 i_2} \ldots p_{i_r i_{r+1}} = \\ &= \sum_{i_1 \ldots i_{r+1}} (d_{i_1} + d_{i_2} + \ldots + d_{i_r}) p_{i_1}(1) p_{i_1 i_2} \ldots p_{i_r i_{r+1}} + \\ &\quad + \sum_{i_1 \ldots i_{r+1}} d_{i_{r+1}} p_{i_1}(1) p_{i_1 i_2} \ldots p_{i_r i_{r+1}} = \\ &= \sum_{i_1 \ldots i_r} (d_{i_1} + d_{i_2} + \ldots + d_{i_r}) p_{i_1}(1) p_{i_1 i_2} \ldots p_{i_{r-1} i_r} + \\ &\quad + \sum_{i=1}^{n} d_i p_i(r+1) = M(\mathfrak{P}^r) + \pi_0 \mathbf{P}^r \mathbf{D}. \end{aligned}$$

By the induction hypothesis, we have

$$\begin{aligned} M(\mathfrak{P}^{r+1}) &= \pi_0 (\mathbf{E} + \mathbf{P} + \ldots + \mathbf{P}^r) \mathbf{D} = \\ &= (r+1)(\pi_0 \mathbf{P}_{r+1} \mathbf{D}). \end{aligned}$$

Thus formula (12) is true for any r.

Doob (1953) proved a (classical) theorem stating that the sequence $\mathbf{P}_1, \mathbf{P}_2, \ldots, \mathbf{P}_r \ldots$ converges to a stochastic matrix $\mathbf{P}'$. The constructive situation is embodied in the following proposition.

Property 7. If

$$\mathbf{P} = \begin{pmatrix} 1-d & d \\ d & 1-d \end{pmatrix}, \quad d \in \{B_g\},$$

then there is no algorithm which, given $\mathbf{P}$, *will produce a natural number* r *such that for any natural numbers* k, l *the matrix* $\mathbf{M}=(a_{ij})$ *defined as* $\mathbf{P}_{r+k}-\mathbf{P}_{r+l}$ *satisfies the condition* $|a_{ij}|<\frac{1}{8}$.

Proof. Let ε and δ be CRNs such that $|\varepsilon| \not> \frac{1}{2}, |\delta| \not> \frac{1}{2}$. Then

$$\begin{pmatrix} \frac{1}{2}+\varepsilon & \frac{1}{2}-\varepsilon \\ \frac{1}{2}-\varepsilon & \frac{1}{2}+\varepsilon \end{pmatrix} \begin{pmatrix} \frac{1}{2}+\delta & \frac{1}{2}-\delta \\ \frac{1}{2}-\delta & \frac{1}{2}+\delta \end{pmatrix} = \begin{pmatrix} \frac{1}{2}+2\varepsilon\delta & \frac{1}{2}-2\varepsilon\delta \\ \frac{1}{2}-2\varepsilon\delta & \frac{1}{2}+2\varepsilon\delta \end{pmatrix}.$$

Thus, if $|\varepsilon|<\frac{1}{2}$ and $|\delta|<\frac{1}{2}$ the matrix $\mathbf{P}^r$ approaches the matrix

$$\begin{pmatrix} \frac{1}{2} & \frac{1}{2} \\ \frac{1}{2} & \frac{1}{2} \end{pmatrix}$$

with increasing r. Hence it follows that

$$\mathbf{P}' = \begin{pmatrix} \frac{1}{2} & \frac{1}{2} \\ \frac{1}{2} & \frac{1}{2} \end{pmatrix}$$

if $d>0$ and $\mathbf{P}'=\mathbf{P}$ if $d=0$. Were an algorithm possible as described, we could determine, given $\mathbf{P}$, a number r such that

$$|p_{12}^{(r)}-p'_{12}| \not> \frac{1}{8}, \tag{13}$$

where $p_{12}^{(r)}$ and p'_{12} are the corresponding elements of the matrices $\mathbf{P}^r$ and $\mathbf{P}'$. Let ϱ be a rational number such that

$$|\varrho-p_{12}^{(r)}|<\frac{1}{8}. \tag{14}$$

It follows from (13) and (14) that

$$\varrho-\frac{1}{4}<p'_{12}<\varrho+\frac{1}{4}. \tag{15}$$

If we now assume that $\varrho\leqslant\frac{1}{4}$ then, by (15),

$$p'_{12}<\frac{1}{2}.$$

In this case, then, $d \not> 0$. On the other hand, if $\varrho > \frac{1}{4}$, then $p_{12}^{(r)} > \frac{1}{8}$. This is not true if $d=0$. Therefore $d > 0$.

Thus the existence of the proposed algorithm is incompatible with the properties of $\{B_g\}$, and so no such algorithm can exist.

NOTES

Dijkman (1961, 1963, 1964) studies simple homogeneous Markov chains from the intuitionistic viewpoint. Certain positive results stated and proved in the language of intuitionism are easily converted into theorems and proofs of the constructive theory of Markov chains, specifically, those concerning properties of ergodic chains.

A wide field of research is devoted to decompositions of stochastic matrices. From the classical standpoint, such problems have been investigated by Davis (1961), Cleave (1962), Chentsov (1968a) and others. But the methods proposed by these authors (with the exception of Davis) have proved unsuitable for effective construction of decompositions outside the field of rational numbers.*

The work of Podnieks (1971a, b) on this topic is also classical in character. However, the ideas of the first-mentioned paper may be used to improve the algorithm presented in §1.

Vasariņš (1971) is already fully constructive. This paper considers the existence of an algorithm which decides whether or not a stochastic $m \times n$ matrix has given decomposition length, and shows that no such algorithm can exist for lengths $k = mn - m + 1$.

Decompositions of doubly stochastic matrices have been studied only sparsely. All that is known to date is that the length of the minimal decomposition of a doubly stochastic $n \times n$ matrix with rational elements is at most $(n-1)^2+1$. We have good grounds to conjecture, however, that in the general constructive case this estimate cannot be proved. Decomposability of doubly stochastic matrices has also been investigated by Birkhoff, von Neumann and Berge (1958).

* The entire field of applicability of these methods is probably exhausted by matrices with rational elements.

Chapter IV

ENUMERABLE PROBABILITY FIELDS

Our constructive probability theory (CPT), which deals with finite spaces of elementary events, is sometimes inadequate to cope with our subject — stochastic automata. We have therefore seen fit to work out the foundations of a CPT for enumerable spaces of elementary events and to include this material in the book. Conceptually, this chapter contains little that is new, since it simply extends the principles set forth in Chapters II and III.

§1. RANDOM EVENTS

Suppose we are given an alphabet $\mathrm{B} = \{\xi_1, \xi_2, \ldots, \xi_l\}$ and a set A of words over B.

Definition 1. We shall say that A is enumerable without repetitions if there exists a normal algorithm $\mathfrak{D}$ over the alphabet $\mathrm{B} \cup \mathrm{A}_0$, applicable to any natural number n, such that

(1) $\forall n (\mathfrak{D}(n) \in A)$;

(2) $\forall m \forall n (m \neq n \supset \mathfrak{D}(m) \,\overline{\underline{\supset}}\, \mathfrak{D}(n))$;

(3) $\forall x \exists n (x \,\overline{\underline{\supset}}\, \mathfrak{D}(n))$, $x \in A$.

An algorithm $\mathfrak{D}$ satisfying these conditions will be called an enumerating algorithm for the set A. It is easy to see that any set which is enumerable without repetitions has infinitely many enumerating algorithms. The following results characterizing sets with this property are presented without proof.

Property 1. *There exist sets which are enumerable without repetitions but not recursive.*

Property 2. *If a set of natural numbers is recursively enumerable and not equal to a finite set, it is enumerable without repetitions.*

Property 3. *If a set A is enumerable without repetitions and has an enumerating algorithm $\mathfrak{D}$ such that*

$$\forall m \forall n (m < n \supset l(\mathfrak{D}(m)) \leqslant l(\mathfrak{D}(n))),$$

then A is recursive ($l(x)$ denotes the number of letters in the word x).

Definition 2. Let E be a set of words over B and $\mathfrak{B}$ a normal algorithm over B, applicable to any word x, $x \in E$. The pair $\Omega = (E, \mathfrak{B})$ will be called a probability field if:

(1) E is enumerable without repetitions;
(2) the values of the algorithm $\mathfrak{B}$ are nonnegative CRNs;
(3) $\sum_{n=0}^{\infty} \mathfrak{B}(\mathfrak{D}(n)) = 1$, where $\mathfrak{D}$ is an enumerating algorithm for E.

From now on we shall call the set E the space of elementary events of Ω; its elements are elementary random events (henceforth we omit the word "random"), denoted by $e_0, e_1, \ldots, e_n, \ldots$. The enumerating algorithm $\mathfrak{D}$ orders the space of elementary events: the first element e_0 is $\mathfrak{D}(0)$, the second, e_1, is $\mathfrak{D}(1)$, and so on. Consequently, any specification of a space of elementary events E will incorporate a definite enumerating algorithm $\mathfrak{D}$. This is done only for convenience, since no essential point is involved. Indeed, if

$$\sum_{n=0}^{\infty} \mathfrak{B}(\mathfrak{D}(n)) = 1$$

and $\mathfrak{D}'$ is another counting algorithm, then also

$$\sum_{n=0}^{\infty} \mathfrak{B}(\mathfrak{D}'(n)) = 1.$$

Definition 3. A (random) event in a PF Ω is any subset* of the space of elementary events E, which is decidable relative to E, i. e., a set of words A satisfying the condition

$$(A \subseteq E) \,\&\, \forall x((x \in E) \supset (x \in A \vee x \notin A)).$$

Simple arguments show that the union and intersection of finitely many events is again an event (union and intersection refer here to the sets which appear here as events). In this context, we must understand the complement of an event A not as the set $\overline{A}$ but rather $\overline{A}E$, though as before we retain the old notation $\overline{A}$ for the complementary event. It is clear that the complement of an event A is indeed an event.

We define two more operations over the field of events. These operations will enable us to clarify a situation characteristic for the CPT developed here. We shall see later, however, that a similar situation obtains in practically any conceivable CPT, provided its point of departure is traditional probability theory.

Let $A_1, \ldots, A_n, \ldots$ be a sequence of events. We denote by $\bigcap_{i=1}^{\infty} A_i$ the set $\{\mathfrak{A}(x) = 0\}$, where the algorithm $\mathfrak{A}$ is defined as follows: applied to a word x, $\mathfrak{A}$ produces the number 1 if $x \notin A_1$; if $x \in A_1$, $x \in A_2$, $\ldots$, $x \in A_n$ but $x \notin A_{n+1}$, then $\mathfrak{A}(x)$ is the CRN defined by the regularly convergent sequence $1, \frac{1}{2}, \ldots, \left(\frac{1}{2}\right)^n, \left(\frac{1}{2}\right)^n, \ldots$; if $x \in A_1$, $x \in A_2$, $\ldots$, $x \in A_n$, $x \in A_{n+1}$, $\ldots$, $\mathfrak{A}(x)$ is the CRN defined by the regularly convergent sequence $1, \frac{1}{2}, \ldots, \left(\frac{1}{2}\right)^n, \left(\frac{1}{2}\right)^{n+1}, \ldots$.

* It should be clear that generalization of our "set" concept does not yield any essentially new properties of random events.

We now define $\bigcup_{i=1}^{\infty} A_i$ to be the set $\{\mathfrak{A}'(x)>0\}$, where $\mathfrak{A}'$ is the algorithm obtained if we interchange the symbols $\in$ and $\notin$ everywhere in the definition of $\mathfrak{A}$.

The properties of the sets $\bigcup_{i=1}^{\infty} A_i$ and $\bigcap_{i=1}^{\infty} A_i$ are characterized by the following proposition.

Proposition 1. Let the space of elementary events E be the set of all natural numbers. Then there exists a sequence of events $A_1, \ldots, A_n, \ldots$ such that neither $\bigcap_{i=1}^{\infty} \overline{A_i}$ nor $\bigcup_{i=1}^{\infty} A_i$ is an event.

Proof. Let $\{R\}$ be a set which is recursively enumerable but not recursive, enumerated by a general recursive function $\varphi(n)$. We define the sets A_i, $i=1, 2, \ldots$, by

$$A_n \approx \{x=\varphi(n-1)\}.$$

Obviously,

$$\bigcup_{i=1}^{\infty} A_i \approx \{R\}.$$

But $\{R\}$ is by assumption not recursive. It is easy to see that the set $\bigcap_{i=1}^{\infty} \overline{A_i}$ is also not recursive. Thus the sequence constructed has the required properties, and this completes the proof.

It follows from Proposition 1 that the events of a PF Ω form neither a Σ-algebra nor a Δ-algebra. It also demonstrates that no reasonable restriction or generalization of the "event" concept can possibly remedy this situation. A more radical approach might lead to a situation in which probabilities of events are noncomputable.

§2. PROBABILITY OF RANDOM EVENTS

We now define the probability of an event and study its properties.

Definition 4. Let $\Omega=(E, \mathfrak{B})$ be a PF and A an event in Ω. Let E_n denote the following finite subset of E:

$$\{x \mathrel{\underline{\underline{\smile}}} e_0 \vee x \mathrel{\underline{\underline{\smile}}} e_1 \vee \ldots \vee x \mathrel{\underline{\underline{\smile}}} e_n\}, \quad n=0, 1 \ldots$$

For each finite set AE_n, we define a CRN s_n as follows: If AE_n is empty, then $s_n=0$; if AE_n is equal to the finite set

$$\{x \mathrel{\underline{\underline{\smile}}} e_{i_1} \vee x \mathrel{\underline{\underline{\smile}}} e_{i_2} \vee \ldots \vee x \mathrel{\underline{\underline{\smile}}} e_{i_k}\},$$

then $s_n=\mathfrak{B}(e_{i_1})+\ldots+\mathfrak{B}(e_{i_k})$. The limit of the sequence of numbers s_n (if it exists) will be called the probability of A, denoted by $p(A)$.

It is easy to see that the sequence $s_0, s_1, \ldots, s_n, \ldots$ is always convergent. Thus every event A has a probability $p(A)$. By Definition 4, an impossible event $\varnothing$ (empty set of elementary events) has probability $p(\varnothing)=0$, while a sure event $A \approx E$ has probability 1. The following properties are readily proved.

Property 4. If A is an event, then

$$p(\bar{A}) = 1 - p(A).$$

Property 5. If the intersection of any two of the events $A_1, \ldots, A_k$ *is an impossible event, then*

$$p(A_1 \cup \ldots \cup A_k) = p(A_1) + \ldots + p(A_k).$$

Property 6. If A is an event contained in an event B, then $p(A) \not> p(B)$.

We now show that the probability measure is countably additive, i. e., possesses a more general additivity property than Property 5.

Property 7. Let $A_1, \ldots, A_n, \ldots$ *be a sequence of pairwise mutually exclusive* events in a PF* Ω, *such that* $A \approx \bigcup_{i=1}^{\infty} A_i$ *is an event in* Ω. *Then*

$$p(A) = \sum_{i=1}^{\infty} p(A_i).$$

Proof. We let c_n, $n=1, 2, \ldots$ denote the numbers $p(A_1) + \ldots + p(A_n)$. We claim that the sequence $c_1, c_2, \ldots, c_n, \ldots$ is convergent; to prove this, we shall exhibit a constructive procedure whereby, given a number ε of the form $\left(\frac{1}{2}\right)^k$, we can find a natural number n such that

$$\forall m (c_{n+m} - c_n < \varepsilon).$$

By Definition 4, given $\varepsilon = \left(\frac{1}{2}\right)^k$ one can find a natural number n' such that

$$\forall m (p(E_{n'+m}) - p(E_{n'}) < \varepsilon). \qquad (1)$$

Considering the set $AE_{n'}$, we now find a natural number $\widehat{n}$ such that

$$AE_{n'} \subseteq \bigcup_{i=1}^{\widehat{n}} A_i. \qquad (2)$$

We claim that $\widehat{n}$ is the required n. Indeed, by (1) and (2) we can write

$$p(A) \not> p(AE_{n'}) + \varepsilon;$$

$$\forall m (p(AE_{n'}) \not> p(A_1 \cup A_2 \cup \ldots \cup A_{\widehat{n}} \cup$$

$$\cup \ldots \cup A_{\widehat{n}+m}) \not> p(A)).$$

* Events A and B are mutually exclusive if $AB \approx \varnothing$.

Hence it follows that

$$\forall m(p(AE_{n'}) \not> c_{\widehat{n}+m} \not> p(AE_{n'}) + \varepsilon).$$

Thus, for all natural numbers m, we have

$$c_{\widehat{n}+m} - c_{\widehat{n}} \not> \varepsilon.$$

It is also easy to see that the limit of the sequence $\{c_n\}$ is precisely $p(A)$.

As a corollary to Property 7, we can prove that the probability measure is continuous.

Property 8. Let $A_1, A_2, \ldots, A_n, \ldots$ be a sequence of events in Ω such that

$$\forall i(A_{i+1} \subseteq A_i);$$

$$\bigcap_{i=1}^{\infty} A_i \approx \varnothing.$$

Then $\lim_i p(A_i) = 0$.

Proof. Construct a new sequence of events $B_1, B_2, \ldots, B_n, \ldots$ by setting $B_n \equiv A_n \overline{A}_{n+1}$. It is easy to see that the events B_n are pairwise mutually exclusive. We shall show that

$$\forall n(\bigcup_{i=n}^{\infty} B_i \approx A_n).$$

Indeed, if e is an elementary event contained in $\bigcup_{i=n}^{\infty} B_i$, then $\mathfrak{A}'(e) > 0$. By the definition of the algorithm $\mathfrak{A}'$, we may thus find a number m for which $e \in B_m$, $m \geqslant n$. Consequently, the event e is contained in A_m. But since A_m is contained in A_n, we have $e \in A_n$. Conversely, if $e \in A_n$ there exists i such that $e \notin A_i$, $i \geqslant n$. Indeed, suppose that this is not true:

$$\neg \exists i((i \geqslant n) \& (e \notin A_i)).$$

Then

$$\forall i((i \geqslant n) \supset \neg \neg (e \in A_i)).$$

But since A_i is an event, we have

$$\forall i((i \geqslant n) \supset (e \in A_i)).$$

Thus the set $\bigcap_{i=1}^{\infty} A_i$ is not empty and this is a contradiction. Consequently,

$$\neg \neg \exists i((i \geqslant n) \& (e \notin A_i)).$$

Since the set A_i is decidable, it follows that

$$\exists i((i \geqslant n) \& (e \notin A_i)).$$

If n' is the first index in the sequence such that $e \notin A_{n'}$, then $e \in B_{n'}$. Consequently, the set $\bigcup_{i=n}^{\infty} B_i$ contains e.

We may now state that

$$p(A_n) = \sum_{i=n}^{\infty} p(B_i).$$

It follows that

$$p(A_n) = p(A_1) - p(B_1 \cup \ldots \cup B_{n-1}) = p(A_1) - \sum_{i=1}^{n-1} p(B_i).$$

Consequently, for any ε, $\varepsilon = \left(\frac{1}{2}\right)^k$, we can find n_ε such that

$$\forall m (p(A_{n_\varepsilon + m}) < \varepsilon).$$

Provided that every elementary event in Ω has positive probability, we can prove two more properties.

Property 9. If $A_1, A_2, \ldots, A_n, \ldots$ is a sequence of events such that $\sum_{i=1}^{\infty} p(A_i)$ is convergent, then $\bigcup_{i=1}^{\infty} A_i$ is an event.

Property 10. If $A_1, \ldots, A_n, \ldots$ is a monotone decreasing sequence of events, i.e., $A_{i+1} \subset A_i$ for all i, and the sequence $p(A_1), \ldots, p(A_n), \ldots$ is convergent, then the set $\bigcap_{i=1}^{\infty} A_i$ is an event.

An important corollary of Property 7 is Bayes' theorem. Before stating the theorem, we define the conditional probability of an event A given H by

$$p(A/H) = \frac{p(AH)}{p(H)}.$$

Of course, this definition is meaningful only if $p(H) > 0$, as we shall indeed assume whenever speaking of conditional probability.

Property 11. Let $H_1, H_2, \ldots, H_n, \ldots$ be a sequence of pairwise mutually exclusive events in Ω, satisfying the conditions:

(1) $\bigcup_{i=1}^{\infty} H_i \approx E$;

(2) $\forall i (p(H_i) > 0)$.

Then, for any event A in Ω,

$$p(A) = \sum_{i=1}^{\infty} p(H_i) p(A/H_i).$$

Proof. It is readily seen that the events in the sequence $AH_1, AH_2, \ldots,$ $\ldots, AH_n, \ldots$ are pairwise mutually exclusive, and $\bigcup_{i=1}^{\infty} AH_i \approx A$. By Property 7, we have

$$p(A) = \sum_{i=0}^{\infty} p(AH_i).$$

But since

$$p(AH_i) = p(H_i)p(A/H_i),$$

it now follows that

$$p(A) = \sum_{i=1}^{\infty} p(H_i)p(A/H_i).$$

§3. RANDOM VARIABLES

The definition of a random variable $\mathfrak{P}$ over a PF Ω is formally the same as that given in Chapter II. The same is true of the notions "sum," "product" and "difference" of random variables. Provided the term "event" is suitably interpreted (see Definition 3), we may also retain Definitions 9 and 10 of Chapter II for independent events and random variables. One must bear in mind that the definition of independent random variables presupposes that all the sets $\{\mathfrak{P}_i = a\}$ are decidable relative to E, where $\mathfrak{P}_i$ runs through all random variables in the system under consideration and a is any of its values.

By contrast, the mean and variance must be defined anew.

Definition 5. The mean of a random variable $\mathfrak{P}$ over Ω is the number

$$M(\mathfrak{P}) = \sum_{i=0}^{\infty} \mathfrak{P}(e_i)\mathfrak{B}(e_i)$$

provided the series $\sum_{i=0}^{\infty} \mathfrak{P}(e_i)\mathfrak{B}(e_i)$ is absolutely convergent.*

We now exhibit a random variable over Ω which has no mean, although any partial sum $\sum_{i=0}^{n} |\mathfrak{P}(e_i)\mathfrak{B}(e_i)|$ of the series is bounded above by 2. Consider a PF for which the probability-measure algorithm is positive, i. e.,

$$\forall i(\mathfrak{B}(e_i) > 0).$$

An example is the algorithm $\mathfrak{B}$ defined by

$$\mathfrak{B}(e_i) = \left(\frac{1}{2}\right)^{i+1}.$$

* Recall that a series $d_1 + \ldots + d_n + \ldots$ is absolutely convergent if the series $|d_1| + \ldots + |d_n| + \ldots$ is convergent.

Now suppose that $\{R\}$ is a set of natural numbers which is enumerable without repetitions by a general recursive function $\varphi(n)$, but not recursive. We define a random variable $\mathfrak{P}$ over Ω by

$$\mathfrak{P}(e_i) = \left(\frac{1}{2}\right)^{\varphi(i)} [\mathfrak{V}(e_i)]^{-1}.$$

It is easy to see that

$$\forall n \left(\sum_{i=0}^{n} \mathfrak{P}(e_i)\mathfrak{V}(e_i) < 2 \right).$$

However, $\mathfrak{P}$ cannot have a mean, for the series in question is not convergent (Rice, 1954).

Definition 6. Let $\mathfrak{P}$ be a random variable over Ω with mean $M(\mathfrak{P})$. The number

$$D(\mathfrak{P}) = \sum_{i=0}^{\infty} (\mathfrak{P}(e_i) - M(\mathfrak{P}))^2\mathfrak{V}(e_i)$$

will be called the variance of $\mathfrak{P}$, provided that the series

$$\sum_{i=0}^{\infty} (\mathfrak{P}(e_i) - M(\mathfrak{P}))^2\mathfrak{V}(e_i)$$

is convergent.

Definition 7. Let $\mathfrak{P}_1$, $\mathfrak{P}_2$ be random variables over Ω which have means. If the product of the random variables $\mathfrak{P}_1 - M(\mathfrak{P}_1)$ and $\mathfrak{P}_2 - M(\mathfrak{P}_2)$ has a mean, we define the covariance of $\mathfrak{P}_1$ and $\mathfrak{P}_2$ as

$$\mathrm{Cov}\,(\mathfrak{P}_1, \mathfrak{P}_2) = M((\mathfrak{P}_1 - M(\mathfrak{P}_1))(\mathfrak{P}_2 - M(\mathfrak{P}_2))).$$

We now scan some important properties of random variables, omitting the simplest proofs.

Property 12. If $\mathfrak{P}_1, \ldots, \mathfrak{P}_n$ are random variables over Ω, all having means, then

$$M(\mathfrak{P}_1 + \ldots + \mathfrak{P}_n) = M(\mathfrak{P}_1) + \ldots + M(\mathfrak{P}_n).$$

Property 13. If a random variable $\mathfrak{P}$ over Ω has a variance, then

$$D(\mathfrak{P}) = M(\mathfrak{P}^2) - (M(\mathfrak{P}))^2.$$

This follows from the fact that the existence of the variance guarantees convergence of the series

$$\sum_{i=0}^{\infty} (\mathfrak{P}(e_i))^2\mathfrak{V}(e_i);$$

$$\sum_{i=0}^{\infty} 2M(\mathfrak{P})\mathfrak{P}(e_i)\mathfrak{V}(e_i).$$

Property 14. If a random variable $\mathfrak{P}$ over Ω has a second moment, i.e., the series

$$\sum_{i=0}^{\infty} (\mathfrak{P}(e_i))^2 \mathfrak{B}(e_i)$$

is convergent, then it has a mean and a variance.

Proof. Let a be some value of $\mathfrak{P}$. Let r be a rational number which approximates $|a|$ to within a positive error ε satisfying the conditions $\varepsilon < \frac{1}{4}$; $|2r\varepsilon + \varepsilon^2| < \frac{1}{4}$.

We have

$$r - \frac{1}{4} < |a| < r + \frac{1}{4};$$

$$r^2 - \frac{1}{4} < a^2 < r^2 + \frac{1}{4}.$$

If $r < 1$, there are two possibilities: $r \leqslant \frac{3}{4}$; $r > \frac{3}{4}$. In the first case,

$$|a| < \frac{3}{4} + \frac{1}{4} = 1 \not> a^2 + 1,$$

and in the second

$$|a| < 1 + \frac{1}{4} < 1 + \left(\frac{3}{4}\right)^2 - \frac{1}{4} < a^2 + 1.$$

Suppose now that $r \geqslant 1$. Then $r \leqslant r^2$, and we get

$$|a| < r + \frac{1}{4} < r^2 + 1 - \frac{1}{4} < a^2 + 1.$$

We have thus shown that for any i, $i = 0, 1, \ldots,$

$$|\mathfrak{P}(e_i)| < (\mathfrak{P}(e_i))^2 + 1.$$

It is now readily shown that the series

$$\sum_{i=0}^{\infty} |\mathfrak{P}(e_i)| \mathfrak{B}(e_i)$$

is convergent.

Thus the random variable $\mathfrak{P}$ has a mean. But since $\mathfrak{P}^2$ by assumption also has a mean, it follows that $\mathfrak{P}$ also has a variance (see Property 2).

Property 15. The covariance of two random variables $\mathfrak{P}_1$, $\mathfrak{P}_2$ over Ω is given by

$$\mathrm{Cov}\,(\mathfrak{P}_1, \mathfrak{P}_2) = M(\mathfrak{P}_1\mathfrak{P}_2) - M(\mathfrak{P}_1)M(\mathfrak{P}_2).$$

Property 16. Let $\mathfrak{P}$, $\mathfrak{P}'$ be independent random variables over Ω, both having a mean; then their product $\mathfrak{P}\mathfrak{P}'$ also has a mean, and

$$M(\mathfrak{P}\mathfrak{P}') = M(\mathfrak{P})M(\mathfrak{P}').$$

Proof. Since $\mathfrak{P}, \mathfrak{P}'$ are independent, we may decide whether any two given values of $\mathfrak{P}$ (or $\mathfrak{P}'$) are equal or distinct. Thus, scanning the first n values of $\mathfrak{P}$, $n=1, 2, \ldots$, i. e., the sequence of numbers $\{\mathfrak{P}(e_0), \mathfrak{P}(e_1), \ldots, \mathfrak{P}(e_n)\}$, we can select all distinct values of $\mathfrak{P}$ and arrange them in a sequence

$$C_n \equiv \{a_1, a_2, \ldots, a_{f_n}\}.$$

Doing the same for $\mathfrak{P}'$, we obtain a sequence

$$C'_n \equiv \{a'_1, a'_2, \ldots, a'_{g_n}\}.$$

Let us assume that the terms in the sequences C_n and C'_n are ordered in accordance with the following rule: If the sequence C_n (C'_n) has already been constructed and it contains no number equal to $\mathfrak{P}(e_{n+1})$ (to $\mathfrak{P}'(e_{n+1})$), this value will appear in the sequences C_{n+m} (C'_{n+m}) after all values appearing in C_n (in C'_n). It is readily understood that the random variable $\mathfrak{P}$ (or $\mathfrak{P}'$) may have the property that all the sequences C_n (or C'_n) coincide for sufficiently large n. Because of this eventuality, we cannot handle the sequences of all distinct values directly for such random variables, since no repetitions are allowed.

We now define a number sequence $c_1, c_2, \ldots, c_n, \ldots$ by

$$c_n = \sum_{i=1}^{f_n} \sum_{j=1}^{g_n} a_i a'_j p(\{\mathfrak{P}=a_i, \mathfrak{P}'=a'_j\}).$$

We claim that this sequence is convergent and its limit c is precisely $M(\mathfrak{P})M(\mathfrak{P}')$.

Indeed, the series

$$\sum_{i=0}^{\infty} |\mathfrak{P}(e_i)|\mathfrak{B}(e_i);$$

$$\sum_{j=0}^{\infty} |\mathfrak{P}'(e_j)|\mathfrak{B}(e_j)$$

are convergent, and therefore so is the sequence $d_1, d_2, \ldots, d_n, \ldots$ defined by

$$d_n = \sum_{i=0}^{n} \mathfrak{P}(e_i)\mathfrak{B}(e_i) \sum_{j=0}^{n} \mathfrak{P}'(e_j)\mathfrak{B}(e_j).$$

Given $\varepsilon=\left(\frac{1}{2}\right)^k$, we can find $n=n_\varepsilon$ such that

$$\forall m\left(|d_{n_\varepsilon+m}-d_{n_\varepsilon}|<\frac{1}{2}\varepsilon\right).$$

The required equality now follows from the fact that

$$|c_{n_\varepsilon}-d_{n_\varepsilon}|=|\sum_{i=0}^{f_{n_\varepsilon}} a_i p(\{\mathfrak{P}=a_i\}) \sum_{j=0}^{g_{n_\varepsilon}} a'_j p(\{\mathfrak{P}'=a'_j\})-d_{n_\varepsilon}|<\varepsilon. \quad (3)$$

We now show that c is the sum of the absolutely convergent series

$$\sum_{i=0}^{\infty} \mathfrak{P}(e_i)\mathfrak{P}'(e_i)\mathfrak{B}(e_i).$$

Slight modifications of the previous arguments readily show that the sequence $\{\widehat{c_n}\}$, where

$$\widehat{c_n}=\sum_{i=0}^{f_n}\sum_{j=0}^{g_n} |a_i a'_j| p(\{\mathfrak{P}=a_i, \mathfrak{P}'=a'_j\}),$$

is convergent.

On the other hand, given n and ε, we can find N such that

$$\forall m\left(\sum_{i=0}^{f_n}\sum_{j=0}^{g_n} |a_i a'_j| (p(\{\mathfrak{P}=a_i, \mathfrak{P}'=a'_j\}) - \right.$$
$$\left. -p(\{\mathfrak{P}=a_i, \mathfrak{P}'=a'_j\}E_{N+m}))<\varepsilon\right). \quad (4)$$

By the convergence of $\{\widehat{c_n}\}$ and inequality (4), we conclude that there exists n' for which

$$\forall m\left(\sum_{i=0}^{f_{n'+m}}\sum_{j=0}^{g_{n'+m}} |a_i a'_j| p(\{\mathfrak{P}=a_i, \mathfrak{P}'=a'_j\}E_{n'+m}) - \right.$$
$$\left. -\sum_{i=0}^{f_{n'}}\sum_{j=0}^{g_{n'}} |a_i a'_j| p(\{\mathfrak{P}=a_i, \mathfrak{P}'=a'_j\}E_{n'})<\varepsilon\right). \quad (5)$$

But since

$$\sum_{i=0}^{f_n}\sum_{j=0}^{g_n} |a_i a'_j| p(\{\mathfrak{P}=a_i, \mathfrak{P}'=a'_j\}E_n) = \sum_{i=0}^{n} |\mathfrak{P}(e_i)\mathfrak{P}'(e_j)|\mathfrak{B}(e_i),$$

it follows from (4) that the sequence $\sum_{i=0}^{\infty} \mathfrak{P}(e_i)\mathfrak{P}'(e_i)\mathfrak{B}(e_i)$ is absolutely convergent (hence, convergent). Our assertion concerning c is now easily proved, in view of the fact that (4) is also valid if the absolute value symbol $|\ |$ is dropped.

As a corollary to Properties 12, 15 and 16, we obtain the Bienaymé equality for enumerable spaces of elementary events:

Property 17. If $\mathfrak{P}_1, \ldots, \mathfrak{P}_n$ is a system of independent random variables over Ω (pairwise independence will suffice), all of which have variances, then their sum also has a variance and

$$D(\mathfrak{P}_1 + \ldots + \mathfrak{P}_n) = D(\mathfrak{P}_1) + \ldots + D(\mathfrak{P}_n).$$

Retaining the interpretation given Chebyshev's inequality in Chapter II, one easily proves it for the case of PFs with enumerable spaces of elementary events.

NOTES

A constructive theory of probability has been developed, independently of the present author, by Kosovskii (1969). His papers are distinguished for the generality of the fundamental definitions and results. His approach is conceptually similar to the classical version of the foundations of probability theory which treats PFs (in Kosovskii's terminology, probability spaces) as normed decidable boolean algebras. Kosovskii holds that an essential component of his constructive conception is the fact that the definitions of "random event" and "random variable" are globally conceived.* We believe, notwithstanding, that as long as one is discussing such simple things as finite and enumerable PFs the definitions of Chapters II through IV provide a simpler and more lucid approach.

It should be noted that we have made no attempt to generalize the random variable concept by using double negations of predicates. Of course, this approach is not without theoretical interest, but it does not seem to offer many practical advantages. We have refrained from the study of specific probability laws such as the Pascal law, Poisson law, etc., since after all our subject is stochastic automata and not probability theory proper. For the same reason, we have not discussed Markov chains with enumerably many states. Here we confine ourselves to an indication of the lines a suitable definition would follow.

Let $\mathfrak{A}$ be a normal algorithm of type N — R and $\mathfrak{B}$ a normal algorithm of type NN — R, satisfying the condition

$$\begin{cases} 0 \not> \mathfrak{A}(n); \\ 0 \not> \mathfrak{B}(m, n); \\ \sum_{n=0}^{\infty} \mathfrak{A}(n) = 1; \\ \forall m \left(\sum_{n=0}^{\infty} \mathfrak{B}(m, n) = 1 \right). \end{cases}$$

A Markov chain will then be a triple $[E, \mathfrak{A}, \mathfrak{B}]$, where E is a set of words over an alphabet B, enumerable without repetitions.

* The importance of global (i.e., nonlocal) definitions of various objects in constructive mathematics has been emphasized particularly by Shanin (1962).

Chapter V

INFINITE STOCHASTIC AUTOMATA

In this chapter we consider various mathematical models of infinite stochastic machines, which we shall call infinite stochastic automata (ISA). We shall study their properties and performance, subject to various interpretations of how they handle the tasks given them. A few questions of constructive probability theory relevant to the basic problems will be touched on in passing.

§1. INTRODUCTORY NOTIONS

We consider two alphabets $\Sigma = \{\sigma_1, \sigma_2, \ldots, \sigma_m\}$ and $\widehat{B} = \{\xi_1, \xi_2, \ldots, \xi_l, \Box\}$.

Let $\mathfrak{B}$ be a normal algorithm over Σ, applicable to any word e over Σ and producing a word over $\widehat{B}$.

We shall assume that $\mathfrak{B}$ satisfies the following conditions:

(1) For no word e is it true that $\Box\Box \mho \mathfrak{B}(e)$. *

(2) If $\mathfrak{B}(e)$ is not the empty word, then its prefix (initial segment) is $\Box\xi_i$ and its suffix (final segment) $\xi_j\Box$.

(3) For any words e and e' over Σ, $\mathfrak{B}(e)$ is a prefix of $\mathfrak{B}(ee')$. This means that $\mathfrak{B}$ transforms any word e into either the empty word or a word of the form $\Box g_1 \Box g_2 \Box \ldots \Box g_n \Box$, where the g_i are nonempty words over B.

Let π_0 denote a stochastic row-vector $(p_1, p_2, \ldots, p_m)$.

Definition 1. A stochastic machine **M** is specified through the system of objects $[\Sigma, \widehat{B}, \mathfrak{B}, \pi_0]$.

Definition 2. A stochastic machine **M** is called a deterministic 1-machine (de Leeuw et al. (1956)) if $p_i = 1$ for some i, $1 \leqslant i \leqslant m$.

There are various generalizations of the concept of a stochastic machine.

Definition 3. A Markov stochastic machine **M** is specified through the system of objects $[\Sigma, \widehat{B}, \mathfrak{B}, \pi_0, \mathbf{P}]$, where **P** is an m-th order stochastic matrix.

Definition 4. An even stronger generalization is obtained if we replace the algorithm $\mathfrak{B}$ by an algorithm $\mathfrak{P}$ with the following properties:

(1) $\mathfrak{P}$ is applicable to any natural number, producing a word $\mathfrak{B}^*_i \langle\rangle p^*_i$, where $\mathfrak{B}^*_i$ is the notation for a normal algorithm satisfying conditions 1 and 2 above and p^*_i a CRN such that $0 \ntriangleright p^*_i \ntriangleright 1$.

* The notation $e \mho f$, where e and f are words over some alphabet, stands for the statement "the word e occurs in the word f."

(2) $\sum_{i=1}^{\infty} p^*_i = 1.$

We shall now try to interpret our definitions of ISA in such a way as to clarify how they model "real" devices. We begin with the concept of stochastic machine as a mathematical model for infinite stochastic machines.

The SM functions as follows. Letters $\sigma_{i_1}, \sigma_{i_2}, \ldots$ are recorded in sequence on a tape, which is then fed into the machine

$$\mathbf{M} = [\Sigma, \widehat{B}, \mathfrak{B}, \pi_0];$$

the words formed by these letters (first letter, first two letters, etc., first k letters) are processed by the algorithm $\mathfrak{B}$, and the results appear on the output tape of the machine $\mathbf{M}$. The letters are distributed on the input tape subject to the probability distribution π_0; that is, σ_i appears as the k-th letter ($k=1, 2, \ldots$) with probability p_i. The output tape first records $\mathfrak{B}(\sigma_{i_1} \sigma_{i_2} \ldots \sigma_{i_k})$, and then the word f defined by

$$\mathfrak{B}(\sigma_{i_1} \sigma_{i_2} \ldots \sigma_{i_k}) f \equiv \mathfrak{B}(\sigma_{i_1} \sigma_{i_2} \ldots \sigma_{i_k} \sigma_{i_{k+1}}).$$

Clearly, if the machine is deterministic, the only letter appearing on its input tape is σ_i (in sequences of increasing length). Consequently, the output (inscription on the output tape) is predetermined; thus, in a deterministic machine one can accurately predict the output tape, without having to activate the machine itself. It is not implied, of course, that this cannot occur in an SM. There are cases (depending on the algorithm $\mathfrak{B}$) when the output tape of a stochastic machine (which is not deterministic in the sense of Definition 2) is fully predetermined. However, one can construct SMs whose output depends on chance and cannot be predicted.

We now proceed to the operation of a Markov stochastic machine (MSM) and its generalizations (see Definition 4). It should be noted that the following interpretation is not the only one possible. Another description may be found in Lorenz (1971).

The difference between an MSM and an SM lies in the composition of the input tape. The first input letter σ_{i_1} appears subject to the distribution π_0, i. e., depending on the result of a game in which the letter σ_i appears with probability p_i, $i=1, 2, \ldots, m$. Suppose now that the k-th input letter is σ_i; then the ($k+1$)-th letter in the sequence is determined by the distribution $\pi_i = (p_{i1}, p_{i2}, \ldots, p_{im})$, where p_{ij}, $j=1, 2, \ldots, m$, are the entries in the i-th row of the stochastic matrix $\mathbf{P}$. The input is then processed exactly as in the stochastic machine. An MSM will function like an SM if $\pi_0 = \pi_i$ for all i, $i=1, 2, \ldots, m$.

In one respect only, the functioning of an ISA

$$\mathbf{M} = [\Sigma, \widehat{B}, \mathfrak{P}, \pi_0, \mathbf{P}]$$

is similar to that of an MSM — the letters of Σ are distributed at the input in the same way. The essential difference lies in the processing of the input data. If the first input letter is σ_{i_1}, the second σ_{i_2}, and so on, and the

k-th is σ_{i_k}, so that the total input to the automaton is the word $\sigma_{i_1}\sigma_{i_2}\ldots\sigma_{i_k}$, the output tape cannot (in the general case) be predicted in advance. The reason is that both the input sequence $\sigma_{i_1}\sigma_{i_2}\ldots\sigma_{i_k}$ and the sequence of algorithms $\mathfrak{B}^*_{j_1}, \mathfrak{B}^*_{j_2}, \ldots, \mathfrak{B}^*_{j_k}$ processing the input data are randomized. The algorithm $\mathfrak{B}^*_i$ will appear as the k-th term of the latter sequence with probability p^*_i, $i=1, 2, \ldots$ We shall assume that appearance of an input sequence $\sigma_{i_1}\sigma_{i_2}\ldots\sigma_{i_k}$ and a sequence of algorithms $\mathfrak{B}^*_{j_1}, \mathfrak{B}^*_{j_2}, \ldots, \mathfrak{B}^*_{j_k}$ produces on the output tape the word obtained from

$$\mathfrak{B}^*_{j_1}(\sigma_{i_1})\mathfrak{B}^*_{j_2}(\sigma_{i_1}\sigma_{i_2})\ldots\mathfrak{B}^*_{j_k}(\sigma_{i_1}\sigma_{i_2}\ldots\sigma_{i_k})$$

by applying a normal algorithm with substitution scheme $\{\Box\Box\rightarrow\Box$.

The following propositions are easily proved.

Proposition 1. There is no algorithm deciding whether or not a given stochastic machine is deterministic.

Proposition 2. There is no algorithm deciding whether or not the output of a stochastic machine is deterministic.

§2. STATEMENT OF THE FUNDAMENTAL PROBLEM

When studying the properties and performance of ISA, it is of paramount importance to delineate the class of functions which they are supposed to implement. We shall first confine attention to stochastic enumeration of sets of natural numbers. At the end of the chapter, however, we shall consider other classes of functions, which may indeed be more interesting. Throughout, the algorithm $\widehat{\mathrm{B}}$ figuring in the definition of an ISA will be the specific alphabet $\widehat{\mathrm{A}}_0=\{0, 1, \Box\}$.

Before formulating the appropriate definitions, we warn the reader that the list of rigorous formulations to be discussed in this chapter is by no means exhaustive. This should be borne in mind when interpreting the results.

Definition 5. Let U be a given set of natural numbers. We shall say that a deterministic machine **M** enumerates U if

$$\forall n(n\in U\leftrightarrow\exists x(\Box n\Box \,\overline{\circ}\, \mathfrak{B}(x))),$$

where x is a word over the alphabet $\{\sigma_i\}$.

Definition 6. Let λ be a CRN, $0<\lambda<1$. We shall say that a stochastic machine M enumerates the set U with cut-point λ if

$$\forall n(n\in U\leftrightarrow\exists k(p_k(n)>\lambda)),$$

where $p_k(n)$ denotes the probability that in k steps **M** will produce a word e containing $\Box n\Box$.

In order to clarify some of the fine points of our definition of stochastic enumerability, let us analyze Definition 6 in detail. With every ISA we may associate a certain sequence of PFs Ω_k, $k=1, 2, \ldots$. To avoid cumbersome notation, we shall retain the symbol Ω_k for the PFs associated with any of the three types of ISA.

Let δ be a letter not contained in the alphabet Σ. With every word $e \boxeq \sigma_{i_1} \sigma_{i_2} \ldots \sigma_{i_k}$ we associate the word $e_\delta \boxeq \delta\sigma_{i_1} \delta\sigma_{i_2} \delta \ldots \delta\sigma_{i_k} \delta$ which we call the δ-image of e. Any word e is clearly uniquely determined by its δ-image.*

For SMs and MSMs, we define the PF Ω_1 as a probability field with space of elementary events

$$E_1 \equiv \{x \boxeq \sigma_1 \vee x \boxeq \sigma_2 \vee \ldots \vee x \boxeq \sigma_m\}$$

and probability measure defined by an algorithm $\mathfrak{B}_1$ such that

$$\forall i ((1 \leqslant i \leqslant m) \supset (\mathfrak{B}_1(\sigma_i) = p_i)).$$

For $k > 1$, we define Ω_k as the field of k-independent trials associated with the PF Ω_1 (in the case of an SM). In other words, the elementary events are the δ-images of words e of length k, and the probability measures $\mathfrak{B}_k$ are defined by

$$\mathfrak{B}_k(\delta\sigma_{i_1} \delta \ldots \delta\sigma_{i_k} \delta) = p_{i_1} p_{i_2} \ldots p_{i_k}.$$

In the case of an MSM, the field Ω_k, $k > 1$, is defined as the field of first Markov k-trials induced by $[E_1, \pi_0, P]$.

The procedure is different for an ISA

$$\mathbf{M} = [\Sigma, \widehat{\mathrm{A}}_0, \mathfrak{P}, \pi_0, \mathbf{P}].$$

We form the PF

$$\Omega_k = (E_k, \mathfrak{B}_k),$$

letting E_k be the set of all words e_δ of the form

$$\delta\sigma_{i_1} \delta \ldots \delta\sigma_{i_k} \delta n_1 \delta \ldots \delta n_k \delta,$$

where n_i are natural numbers, and

$$\mathfrak{B}_k(e_\delta) = p_{i_1} p_{i_1 i_2} \ldots p_{i_{k-1} i_k} p^*_{n_1} p^*_{n_2} \ldots p^*_{n_k}.$$

E_k is obviously enumerated without repetitions by some algorithm; the trivial proof of this statement is omitted. It is also easily proved that the algorithm $\mathfrak{B}_k$ does indeed define a probability measure over E_k.

We now define a random variable $\mathfrak{P}_{nk}$ over Ω_k. For ISA of the first and second types (SM and MSM), the definition is identical, so we shall not specify whether the algorithm $\mathfrak{B}$ operates in an SM or an MSM.

Thus, let Ω_k be the probability field in question. Then $(\mathfrak{P}_{n1}\sigma_i) = 1$ if $\Box n \Box \mho\, \mathfrak{B}(\sigma_i)$; otherwise, $\mathfrak{P}_{n1}(\sigma_i) = 0$.

If $k > 1$, then $\mathfrak{P}_{nk}(e_\delta) = 1$ provided $\Box n \Box \mho\, \mathfrak{B}(e)$; otherwise, $\mathfrak{P}_{nk}(e_\delta) = 0$.

Now let Ω_k be associated with an ISA of the third type, i. e.,

* Our introduction of δ-images at this point is dictated by considerations of uniformity. The idea is to stress the connection between the PF Ω_k and such notions as the "field of r-independent trials associated with a probability field Ω" and the "field of first Markov r-trials induced by a Markov chain $[E, \pi_0, P]$."

$$\mathbf{M}=[\Sigma, \widehat{A}_0, \mathfrak{P}, \pi_0, \mathbf{P}].$$

We define $\mathfrak{P}_{nk}$ over Ω as follows:

$$\mathfrak{P}_{nk}(\delta\sigma_{i_1}\delta\ldots\delta\sigma_{i_k}\delta n_1\delta\ldots\delta n_k\delta)=1,$$

if

$$\square n\square \mho \mathfrak{B}_{n_1}(\sigma_{i_1})\mathfrak{B}_{n_2}(\sigma_{i_1}\sigma_{i_2})\ldots\mathfrak{B}_{n_k}(\sigma_{i_1}\sigma_{i_2}\ldots\sigma_{i_k});$$

otherwise,

$$\mathfrak{P}_{nk}(\delta\sigma_{i_1}\delta\ldots\delta\sigma_{i_k}\delta n_1\delta\ldots\delta n_k\delta)=0.$$

We can now give a rigorous expression for the probability $p_k(n)$ figuring in Definition 6:

$$p_k(n)=p(\{\mathfrak{P}_{nk}=1\}).$$

It is readily seen that

$$\forall n\forall k(M(\mathfrak{P}_{nk})=p_k(n)).$$

Fixing n, let us consider the final mean $M(\mathfrak{P}_n)$. If the sequence

$$M(\mathfrak{P}_{n1}), M(\mathfrak{P}_{n2}), \ldots, M(\mathfrak{P}_{nk}), \ldots$$

is indeed convergent, we might call its limit $M(\mathfrak{P}_n)$ the probability that the ISA **M** will produce the number n on its output tape. This approach (which is that of de Leeuw et al. (1956)) is quite legitimate from the classical standpoint, since the sequence is monotone nondecreasing and bounded above by 1. In the constructive context, however, some care must be taken. Highly instructive in this respect is the result of Specker (1949) concerning the infimum of a monotone increasing bounded sequence of rational numbers. As a justification of Definition 6, we shall construct an SM **M** in which one cannot speak of the probability that $\widehat{\mathbf{M}}$ will produce a number n.

Let Σ consist of only two letters σ_1 and σ_2 $(n=2)$. Using a nonrecursive recursively enumerable set of natural numbers $\{Q\}$, we define an algorithm $\mathfrak{B}$ over the alphabet $\Sigma\cup\widehat{A}_0$. To simplify matters, we assume that the number 0 is not in $\{Q\}$. Let $\varphi(n)$ be a general recursive function enumerating $\{Q\}$ without repetitions; then the algorithm $\mathfrak{B}$ is defined as follows: if $l(e)=k$, we set $\mathfrak{B}(e)\mho\square 0|\square$ if $\nu(e)\leqslant c_k=2^k\times\left[\left(\frac{1}{2}\right)^{n_1}+\left(\frac{1}{2}\right)^{n_2}+\ldots+\left(\frac{1}{2}\right)^{n_l}\right]$; $\mathfrak{B}(e)\mho$ (i. e., the empty word) otherwise.

Here $\nu(e)$ denotes the position of the word e in the lexicographic ordering of all words of length k over Σ, and the natural numbers $n_1, n_2, \ldots, n_l$ exhaust all the elements of the set

$$\{(x=\varphi(0)\vee x=\varphi(1)\vee\ldots\vee x=\varphi(k-1))\&(x\leqslant k)\}.$$

If $\forall x((x\leqslant k)\supset(\varphi(x)>k))$, we define $c_k=0$. Clearly, this algorithm $\mathfrak{B}$ satisfies all the conditions required by the definition of an SM. Our SM $\widehat{M}$ is now specified through the alphabets Σ and $\widehat{A}$, the algorithm $\mathfrak{B}$ and the stochastic vector $\pi_0 = \left(\frac{1}{2}, \frac{1}{2}\right)$.

It remains only to demonstrate that $\widehat{M}$ possesses the desired properties. To this end, let us determine the probability $p_k(1)$. By the definition of Ω_k, we have

$$\forall e_\delta \left(\mathfrak{B}_k(e_\delta) = \left(\frac{1}{2}\right)^k \right), \qquad e_\delta \in E_k.$$

This means that

$$p_k(1) = p(\{\mathfrak{P}_{1k} = 1\}) = \frac{c_k}{2^k}.$$

Consequently, the numbers $M(\mathfrak{P}_{1k})$ are partial sums of the series

$$\sum_{i=0}^{\infty} \left(\frac{1}{2}\right)^{\varphi(i)}.$$

But this series is divergent, and so we cannot speak of the final mean $M(\mathfrak{P}_1)$.

It should be noted that the machine $\widehat{M}$ is so defined that its interpretation as a p-machine with an effective real number p does not dictate the extension of the effective real number concept used by de Leeuw et al. This counters the objection that our example proves nothing decisive for de Leeuw and his co-authors, since their conception of an effective real number is utterly different from ours.

§3. FUNDAMENTAL EQUIVALENCES

To assist the reader's understanding of the sequel, it is important to clear up one more point concerning the functioning of an ISA. Thanks to the way we have defined the distribution of data on the output tape of an SM, we can always say that the empty word is produced there. For example, if the output word is $\square n_1 \square n_2 \square \ldots$, we may say equally well that first the empty word is produced, and then the word $\square n_1 \square n_2 \square \ldots$.

Put differently, suppose that the SM produces the word $\square n_1 \square n_2 \square \ldots \square n_k \square$ on the output tape, and then stops producing data. Now this actually means that machine continues to produce empty words on the output tape. Finally, the SM may resume production of nonempty words after producing only empty words for some time. We may thus assert: Any recursive set of natural numbers is enumerable by some DM.

We now prove two lemmas from which all the important results of this chapter will follow.

Lemma 1. If a set U of natural numbers is enumerable by some ISA M, then

$$U \cup \{x \equiv \quad\} \approx \{\mathfrak{B}^*(x) > 0\},$$

where x is a variable for words over U.

Proof. We first stipulate that the algorithm $\mathfrak{B}^*$ transforms the empty word into the number 1. To determine what $\mathfrak{B}^*$ does to other words, we define the k-th approximation of a CRN a to be the term r_m in a sequence of rational numbers $r_1, r_2, \ldots, r_m, \ldots$ defining a such that

$$\forall l\left(|r_{m+l} - r_m| < \left(\frac{1}{2}\right)^k\right).$$

We now construct $\mathfrak{B}^*$ by the following rules: $\mathfrak{B}^*(n)$ is the CRN defined by the pair of normal algorithms $\mathfrak{A}_{1n}$, $\mathfrak{A}_{2n}$ such that

$$\forall k(\mathfrak{A}_{2n}(k) = k);$$

$$\mathfrak{A}_{1n}(k) = \left(\frac{1}{2}\right)^k$$

for $k=0, 1, 2$, the procedure continuing as long as the k-th approximation of the number $p_k(n) - \lambda$ does not exceed $\left(\frac{1}{2}\right)^k$; if the $k+1$-th approximation of $p_{k+1}(n) - \lambda$ is greater than $\left(\frac{1}{2}\right)^{k+1}$, we put

$$\forall\ l\left(\mathfrak{A}_{1n}(k+l) = \left(\frac{1}{2}\right)^k\right).$$

It is readily shown that

$$\forall n(\exists k(p_k(n) > \lambda) \longleftrightarrow \mathfrak{B}^*(n) > 0)).$$

This proves Lemma 1.

Lemma 2. Any nonempty set $\{\mathfrak{P}(x) > a\}$ of words over B *is enumerable.*

Proof. Suppose that all words over B have been enumerated (without repetitions) in such a way that the first word in the sequence $e_0, e_1, \ldots, e_n, \ldots$, i. e., e_0, is known to belong to $\{\mathfrak{P}(x) > a\}$. To simplify the notation, we let a_n denote $\mathfrak{P}(e_n) - a$ and $r_n(k)$ the k-th approximation of this number. The enumerating algorithm $\mathfrak{D}$ is defined as follows.

The algorithm $\mathfrak{D}$ transforms 0 into e_0. For a natural number $n > 0$, the algorithm performs several steps. It first transforms n into the word $a_1 \square a_2 \square \ldots \square a_n$, and then into the word

$$\mathfrak{D}(0) \square \mathfrak{D}(1) \square \ldots \square \mathfrak{D}(n-1) \square\square r_1(n) \square r_2(n) \square \ldots \square r_n(n).$$

The next step produces a word

$$\alpha_1 \square \ldots \square \alpha_n \square\square \beta_1 \square \ldots \square \beta_n,$$

where α_i (β_i) is either 0 or 1, the following conditions being satisfied:

(1) $\alpha_i = 1$ if and only if there exists n', $0 \leqslant n' \leqslant n-1$, such that $\mathfrak{D}(n') = e_i$.

(2) $\beta_i = 1$ if and only if $r_i(n) > \left(\frac{1}{2}\right)^n$.

The algorithm $\mathfrak{D}$ then proceeds as follows: if $\beta_i \leqslant \alpha_i$ for all i, $1 \leqslant i \leqslant n$, then $\mathfrak{D}(n) \overline{\overline{\frown}} e_0$; if there exists i such that $\beta_i > \alpha_i$, the algorithm finds the first such i, say $\widehat{i}$, and then $\mathfrak{D}(n) \overline{\overline{\frown}} e_{\widehat{i}}$.

It is readily seen that $\mathfrak{D}$ is so designed that all its values are words in the set $\{\mathfrak{P}(x) > a\}$. Conversely, let e_m be a word in this set. Then for some k we have $r_m(k) > \left(\frac{1}{2}\right)^k$. If k' is the smallest k (computed according to the convergence regulator) for which this inequality holds, the algorithm $\mathfrak{D}$, applied to the number $n' = \max(k', m)$, produces the intermediate result

$$\alpha_1 \square \ldots \square \alpha_{n'} \square\square \beta_1 \square \ldots \square \beta_{n'},$$

where $\alpha_m = 0$, $\beta_m = 1$. Thus one of the values

$$\mathfrak{D}(n'),\ \mathfrak{D}(n'+1),\ \ldots,\ \mathfrak{D}(2n'-1)$$

is graphically equal to e_m. This completes the proof of Lemma 2.

As a corollary to Lemmas 1 and 2, we obtain the following

T h e o r e m 1. *If U is a set of natural numbers enumerable by an ISA with cut-point λ, then U is enumerable by a DM.*

It is easy to see that Theorem 1 may be generalized. We need only replace the alphabet $\widehat{A}_0$ by an arbitrary alphabet $\widehat{B}$, suitably modifying the assumptions of the theorem.

§4. COMPUTABLE QUASISEQUENCES OF WORDS

Hitherto we have understood the word "sequence" in the sense of an algorithm applicable to any natural number n and producing some word over the alphabet B. The specific features of the problem with which we wish to deal require a generalization of this concept.

D e f i n i t i o n 7. Let $\{Q\}$ be a given set of words over the alphabet $\widehat{B}_0$. We shall call $\{Q\}$ a quasisequence of words over B if it satisfies the following conditions for any natural number k:

(1) $\square e_1 \square \ldots \square e_k \square \,\overline{\overline{\not\frown}}\, \square \widehat{e_1} \square \ldots \square \widehat{e_k} \square \supset \neg (Q(\square e_1 \square \ldots \square e_k \square) \,\&\, Q(\square \widehat{e_1} \square \ldots \square \widehat{e_k} \square))$;

(2) $\forall k \neg \forall x_k (\neg Q(x_k))$,

where x_k is a word of type $\square e_1 \square \ldots \square e_k \square$;

(3) $Q(\square e_1 \square \ldots \square e_k \square) \supset Q(\square e_1 \square \ldots \square e_{k-1} \square)$

D e f i n i t i o n 8. Let $\{Q\}$ be a quasisequence of words over B. We shall say that a DM c o m p u t e s $\{Q\}$ if, for any k,

$$(Q(\square e_1 \square \ldots \square e_k \square) \longleftrightarrow \exists x (\mathfrak{B}(x) \overline{\overline{\frown}} \square e_1 \square \ldots \square e_k \square e_{k+1} \square \ldots \square e_{k+l} \square)).\ *$$

* The word x consists of the unique letter σ_i for which $p_i = 1$.

Definition 9. We shall say that an ISA

$$\mathbf{M}=[\Sigma, \widehat{\mathrm{B}}, \mathfrak{B}, \pi_0, \mathbf{P}]$$

or

$$\mathbf{M}=[\Sigma, \widehat{\mathrm{B}}, \mathfrak{P}, \pi_0, \mathbf{P}]$$

computes a quasisequence $\{Q\}$ over B with cut-point λ if

$$\forall x(Q(x) \longleftrightarrow \exists k(p_k(x) > \lambda)),$$

where x is a variable for words of the form $\square e_1 \square \ldots \square e_k \square$ over B and $p_k(x)$ is the probability that, in response to an input word e, $l(e)=k$, the automaton will produce an output word with prefix x.

It is clear that this probability $p_k(x)$ may be given a rigorous meaning in terms of the PFs Ω_k associated with our ISA. It is also readily seen that for every value of x

$$\forall k(p_k(x) \not> p_{k+1}(x)).$$

This implies that the following lemma is valid for quasisequences.

Lemma 1'. If a quasisequence $\{Q\}$ *of words over* B *is computable with cut-point* λ *by some ISA* M, *then the set of words* $\{Q\} \cup \{x \,\underline{\underline{\circ}}\quad\}$ *is equal to the set* $\{\mathfrak{B}'(x) > 0\}$,

As a direct corollary of Lemmas 1' and 2, we obtain

Theorem 2. Any quasisequence $\{Q\}$ *of words over* B *which is computable with cut-point* λ *by some ISA* M *is also computable by a suitable DM.*

NOTES

The concepts of SM and DM were defined in de Leeuw et al. (1956). These authors, however, base their definition on an entirely different concept of CRN. For them, a real number (in the classical sense) is effectively computable if its decimal notation is computable by a suitable Turing machine. The main drawback of this definition (not to speak of the high proportion of classicism it still involves) is that the set of all such numbers is not closed under addition. Because of this, our definition of the symbol $p_k(n)$ is untenable if we use de Leeuw's definition of CRN. There may, however, be a justification for his definition in the semiconstructive approach suggested by Uspenskii (1960). According to the latter's conceptions, the set of CRNs, defined as in the famous papers of Turing (1936, 1937) (de Leeuw et al. are close to this approach), is closed under addition and multiplication. However, this is possible only by dint of Uspenskii's not fully constructive treatment of the existential quantifier.

De Leeuw et al. also present semiconstructive prototypes of several other notions considered in this chapter. For example, our concept of a set of natural numbers enumerable with cut-point λ by an SM corresponds to de Leeuw's concept of a set (weakly) enumerable by p-machines, i. e., machines $\mathbf{M}$ with cut-point $\lambda=\frac{1}{2}$, in which $\Sigma=(1, 0)$ and $\pi_0=(p, 1-p)$. The de Leeuw analog of our quasisequences computable with cut-point λ is the (nonconstructive) concept of a sequence of signs (weakly) computable by a p-machine.

A slight modification of Definition 9 (enumerability of quasisequences with cut-point λ) leads to a reasonable analog of the nonconstructive notion that de Leeuw et al. call strong enumerability of a set. The same holds for their concept of a_p-machines — the corresponding object in our approach would be a machine whose input tape is the dyadic expansion of a CRN p, $0<p<1$, computed by a suitable algorithm. Under our interpretation of CRNs, a machine of this type cannot be specified for every p.

The original reason for the interest in properties of ISA was the general impression that a constructive interpretation of the stochastic machine concepts of de Leeuw et al., based on a more general definition of CRN, would in some way reveal an essential advantage of SMs over DMs. These hopes are defeated by Theorems 1 and 2, which are quite surprising, especially in view of the main theorem of de Leeuw et al., which runs as follows: A set U (sequence A) is enumerable (computable) by some p-machine if and only if it is enumerable (computable) by an a_p-machine.

It seems to us that Lemmas 1, 1' and 2 reveal the mechanism underlying these negative results. They also demonstrate that the very conception of just what one expects of a stochastic machine to do is in need of radical modification.

Chapter VI

ANALYSIS OF FINITE STOCHASTIC AUTOMATA

The material of this chapter is highly diversified. The only thing common to the problems considered is their logical structure, which as a rule does not involve construction of an automaton with specific properties. Instead, we wish to ascertain what additional properties finite stochastic automata (FSA) will have when they satisfy various conditions. An exception to this general rule is provided by problems in which we wish to construct algorithms deciding whether or not FSA in a given class have some property. Such problems nevertheless arise quite naturally in the general context of analysis of the properties presented by FSA.

§1. BASIC NOTIONS AND NOTATION

Definition 1. Let $\Sigma=\{\sigma_1, \sigma_2, \ldots, \sigma_m\}$ and $S=\{s_1, s_2, \ldots, s_n\}$ be two ordered alphabets, $M_m{}^n=\{\mathbf{M}_{\sigma_1}, \mathbf{M}_{\sigma_2}, \ldots, \mathbf{M}_{\sigma_m}\}$ a system of stochastic $n\times n$ matrices, $\pi_0=(p_1, p_2, \ldots, p_n)$ a stochastic vector, and $\alpha=\begin{pmatrix}\alpha_1\\ \alpha_2\\ \ldots\\ \alpha_n\end{pmatrix}$ a column vector, where

$$\forall i((\alpha_i=0)\vee(\alpha_i=1)), \quad 1\leqslant i\leqslant n.$$

The system $\mathcal{A}=[\Sigma, S, M_m{}^n, \pi_0, \alpha]$ will be called a finite stochastic automaton.

The alphabets Σ and S are known respectively as the input state alphabet and internal state alphabet of $\mathcal{A}$. The matrices $\mathbf{M}_{\sigma_i}$, $i=1,2,\ldots,n$, will be called the transition matrices of the automaton, and π_0 the initial distribution of internal states. The column vector α defines the set of final internal states of $\mathcal{A}$ and is called the final state vector.

Let $x\equiv\sigma_{i_1}\sigma_{i_2}\ldots\sigma_{i_k}$, $k\geqslant 1$, be a word over Σ. We shall use the notation $\mathbf{M}_x$ for the following product of stochastic matrices:

$$\mathbf{M}_{\sigma_{i_1}}\mathbf{M}_{\sigma_{i_2}}\ldots\mathbf{M}_{\sigma_{i_k}}.$$

The symbol $p_{\mathcal{A}}(x)$ will stand for the number defined by

$$p_{\mathcal{A}}(x)=\pi_0\mathbf{M}_x\alpha.$$

If x is the empty word, then $p_{\mathscr{A}}(x)=\pi_0\alpha$. The (i, j)-th entry of $\mathbf{M}_x$, $1\leqslant i, j\leqslant n$, will be denoted by $a_{ij}(x)$. If x is the empty word, we set $a_{ii}(x)=1$ and $a_{ij}(x)=0$ for $i\neq j$.

According to the customary interpretation of the components of the FSA $\mathscr{A}$ and the above symbols, the entry $a_{ij}(\sigma_k)$ of the matrix $\mathbf{M}_{\sigma_k}$ is the probability that the automaton will go from state s_i to state s_j in response to input σ_k. A similar interpretation holds for the entry $a_{ij}(x)$ of $\mathbf{M}_x$: it characterizes the probability that $\mathscr{A}$ will go from state s_i to state s_j in response to an input sequence $\sigma_{i_1}, \sigma_{i_2}, \ldots, \sigma_{i_r}$, on the assumption that the transition probabilities at the k-th state change of the automaton depend only on the corresponding input signal. On this basis, $p_{\mathscr{A}}(x)$ is the probability that the automaton will reach a final state if the input sequence is $\sigma_{i_1}, \sigma_{i_2}, \ldots, \sigma_{i_r}$ and the initial state is s_i with probability p_i.

This interpretation is not essential as far as the abstract theory of FSA is concerned; it acquires significance only when one is interested in practical applications. An interesting discussion of this aspect of the theory may be found in Ashby (1958). We shall not consider these questions here.

Superficially, the constructive definition of an FSA is no different from the classical definition of Paz (1966a).

Nonetheless, the constructive model is quite different from its classical analog, for the entries of the stochastic matrices $\mathbf{M}_{\sigma_i}$ and the components of the vectors π_0 and α must now be constructive real numbers, i. e., regularly convergent constructive sequences of rational numbers (Markov) or real duplexes (Shanin).

We shall need some additional notation and definitions. Let λ be a CRN, $0<\lambda<1$. We denote by $T(\mathscr{A}, \lambda)$ the set of words x over the input state alphabet Σ of $\mathscr{A}$ defined by the predicate $p_{\mathscr{A}}(x)>\lambda$.

Definition 2. We shall say that the FSA $\mathscr{A}$ *represents* a set of words U over Σ *with cut-point* λ if $U\approx T(\mathscr{A}, \lambda)$.

A set of words representable by an FSA with cut-point λ is often called an *event*, and we shall adopt this usage. One of Rabin's most successful innovations was the concept of an isolated cut-point and the idea of studying the events representable by an FSA with isolated cut-point.

As far as we know, the designation "cut-point" made its first appearance in Rabin (1963a), although the concept itself and the notion of sets of symbols representable with a given cut-point may be traced in the earlier paper of de Leeuw et al. (1956). The difference is that de Leeuw and his co-authors consider stochastic machines and not FSA, but in our opinion this does not change the situation.

Definition 3. Let $\mathscr{A}$ be an FSA. We shall say that $\mathscr{A}$ represents an event U with *isolated cut-point* λ if $U\approx T(\mathscr{A}, \lambda)$ and, for some positive rational number ε, the inequality

$$|p_{\mathscr{A}}(x)-\lambda| \nless \varepsilon$$

holds for any word x over Σ. The number λ will be called an isolated cut-point for the automaton.

§2. QUASIDEFINITE SYSTEMS OF STOCHASTIC MATRICES

Definition 4. A system of matrices $M_m{}^n$ is said to be *quasidefinite* if there is a natural number μ such that any matrix $\mathbf{M}_x$, $l(x)=\mu$, has at least one positive column.

Theorem 1. Suppose that for every real number a appearing as an entry in the matrix $\mathbf{M}_{\sigma_i}$

$$(a=0)\vee(a\neq 0); \qquad (1)$$

then there is an effective procedure telling whether or not the system $M_m{}^n$ *is quasidefinite.*

Proof. For every matrix $\mathbf{M}_x$, we define a new matrix $\mathbf{M}'_x$ by

$$a'_{jk}(x)=\operatorname{sign}(a_{jk}(x)).$$

We now define an operation $\otimes$, which we call logical multiplication, for matrices $\mathbf{C}=(c_{ij})$, $\mathbf{D}=(d_{ij})$ satisfying condition (1), setting $\mathbf{C}\otimes\mathbf{D}=\mathbf{M}$, where the entries a_{ij} of $\mathbf{M}$ are defined by

$$a_{ij}=\operatorname{sign}\sum_{k=1}^{n} c_{ik}d_{kj}.$$

It follows from the definitions that

$$\mathbf{M}'_{x\widehat{x}}=\mathbf{M}'_x\otimes\mathbf{M}'_{\widehat{x}}.$$

This means that $M_m{}^n$ is a quasidefinite system if and only if there is a natural number μ such that any matrix

$$\mathbf{D}'=\mathbf{M}'_{\sigma_{i_1}}\otimes\mathbf{M}'_{\sigma_{i_2}}\otimes\ldots\otimes\mathbf{M}'_{\sigma_{i_\mu}}$$

contains a column of ones. We shall call any number μ satisfying this condition a *quasidefinite index* of $M_m{}^n$.

We claim that if $M_m{}^n$ is quasidefinite it must have a quasidefinite index satisfying

$$\mu<2^{n^2}. \qquad (2)$$

Suppose the contrary, i. e., all quasidefinite indices of $M_m{}^n$ are greater than $2^{n^2}-1$. By assumption, there exists a sequence of matrices $\mathbf{M}_{\sigma_{i_1}}$, $\mathbf{M}_{\sigma_{i_2}}$, $\ldots$, $\mathbf{M}_{\sigma_{i_{2^{n^2}-1}}}$, such that none of the matrices $\mathbf{D}_1, \mathbf{D}_2, \ldots, \mathbf{D}_{2^{n^2}-1}$, where

$$\mathbf{D}_j=\mathbf{M}'_{\sigma_{i_1}}\otimes\ldots\otimes\mathbf{M}'_{\sigma_{i_j}},$$

contains a column of ones. Then not all the matrices $\mathbf{D}_j$ are distinct, and so, for some pair of matrices

$$\mathbf{D}_k,\ \mathbf{D}_l,\qquad 1\leqslant k<l\leqslant 2^{n^2}-1,$$

we have $\mathbf{D}_k=\mathbf{D}_l$.

Using this equality and the definition of the matrices $\mathbf{D}_j$, we get

$$\mathbf{D}_k=\mathbf{D}_k\otimes\{\mathbf{M}'_{\sigma_{i_{k+1}}}\otimes\ldots\otimes\mathbf{M}'_{\sigma_{i_l}}\}=\mathbf{D}_k\otimes\{\mathbf{M}'_{\sigma_{i_{k+1}}}\otimes\ldots\otimes\mathbf{M}'_{\sigma_{i_l}}\}^h,$$

where h is an arbitrary natural number. But this contradicts the assumption that the system is quasidefinite.

We have thus shown that the assumption that a quasidefinite system $M_m{}^n$ has no quasidefinite index less than 2^{n^2} yields a contradiction. Now, for every μ, we may effectively decide whether or not it is a quasidefinite index of $M_m{}^n$; thus we have an effective procedure which, given a system satisfying condition (1), decides whether or not it is quasidefinite.

We may now ask: does there exist an algorithm which, given any system of stochastic $n\times n$ matrices [i. e., not necessarily satisfying (1)], decides whether or not it is quasidefinite? The answer is in the negative:

There exists no algorithm deciding for any given system of stochastic $n\times n$ matrices whether or not it is quasidefinite.

This follows from Property 4 in Chap. III, §1, since if $M_1{}^n$ is quasidefinite, then the matrix $\mathbf{M}_{\sigma_1}$ is regular.

Given any FSA $\mathcal{A}$, one may ask whether its internal states s_i are accessible (Paz, 1966a). In other words, given an internal state s_i we wish to find a word x over Σ such that the i-th component of the stochastic vector $\pi_0(x)=\pi_0\mathbf{M}_x$ is positive. It turns out that this problem is unsolvable even for FSA with quasidefinite systems $M_m{}^n$.

Indeed, consider the accessibility problem for the FSA

$$\mathcal{A}=[\Sigma,\ S,\ M_1{}^2,\ \pi_0,\ \alpha],$$

where

$$\mathbf{M}_{\sigma_1}=\begin{pmatrix}1-d & d\\ 1-d & d\end{pmatrix},\qquad d\in\{B_g\}.$$

It is clear that $\mathbf{M}_x=\mathbf{M}_{\sigma_1}$. Consequently, the accessibility problem for the internal state s_2, when $\pi_0=(1,\ 0)$, is equivalent to the decision problem for the formula $((d=0)\vee(d\neq 0))$, which is unsolvable.

§3. MULTIPLICATIVE PROPERTIES OF QUASIDEFINITE SYSTEMS

Multiplicative properties of stochastic matrices are important not only in the theory of FSA but also for analysis of finite Markov chains. Although some authors have expressed doubts as to the power of matrix methods, it seems to us that from the practical standpoint the approach has fully justified itself. It is unfortunate that little attention has been devoted to numerical methods for approximating the entries of stochastic matrices which are products of other stochastic matrices. The analog-statistical method of Boichenko and Gladkii (1965a, b) seems to us to be successful in its methodological approach. From the same standpoint, the computational scheme proposed by Rabin (1963), based on computer simulation of certain

FSA, is also worthy of attention. Practical application of this scheme, however, involves considerable difficulties, not least of which is the lack of an effective procedure to construct the necessary isolated cut-points.

We are going to prove a constructive analog of a theorem proved in Paz (1965), which plays a major role not only in the theory of FSA but also in the theory of finite Markov chains.

We first define a special symbol $\varkappa(\mathbf{C})$ for an arbitrary matrix $\mathbf{C}$ whose entries are CRNs.

If $\mathbf{C}$ is a column matrix,

$$\mathbf{C}=\begin{pmatrix} c_1 \\ c_2 \\ \dots \\ c_n \end{pmatrix},$$

we set

$$\varkappa(\mathbf{C})=\max(c_1, c_2, \dots, c_n)-\min(c_1, c_2, \dots, c_n);$$

if $\mathbf{C}$ has m columns $\mathbf{B}_1, \mathbf{B}_2, \dots, \mathbf{B}_m$, then

$$\varkappa(\mathbf{C})=\max(\varkappa(\mathbf{B}_1), \varkappa(\mathbf{B}_2), \dots, \varkappa(\mathbf{B}_m)).$$

Lemma 1. Let $\mathbf{C}$ *be a stochastic* $n\times n$ *matrix with positive* k*-th column*

$$\mathbf{B}_k=\begin{pmatrix} c_{1k} \\ c_{2k} \\ \dots \\ c_{nk} \end{pmatrix}$$

and $\mathbf{Q}$ *a column matrix,*

$$\mathbf{Q}=\begin{pmatrix} q_1 \\ q_2 \\ \dots \\ q_n \end{pmatrix},$$

such that $0 \not> q_i \not> 1$ *for every* $i, i=1, 2, \dots, n$. *Then*

$$\varkappa(\mathbf{CQ}) \not> (1-\delta_k)\varkappa(\mathbf{Q}), \tag{3}$$

where

$$\delta_k=\min(c_{1k}, c_{2k}, \dots, c_{nk}). \tag{4}$$

Proof. Denote the entries of the matrix $\mathbf{CQ}$ by q^*_i, $i=1, 2, \dots, n$; we shall determine an upper bound for the differences $q^*_i-q^*_j$, $1\leqslant i, j\leqslant n$. By the definition of matrix multiplication, we have

$$q^*_i = \sum_{t=1}^{n} c_{it} q_t;$$

$$q^*_j = \sum_{t=1}^{n} c_{jt} q_t.$$

Let c denote the number $\max(q_1, q_2, \ldots, q_n)$, and d the number $\min(q_1, q_2, \ldots, q_n)$. Then it follows from the equalities

$$c_{ik} q_k = c c_{ik} - (c - q_k) c_{ik};$$
$$c_{jk} q_k = d c_{jk} + (q_k - d) c_{jk}$$

that

$$q^*_i = \sum_{t=1}^{k-1} c_{it} q_t + c c_{ik} - (c - q_k) c_{ik} + \sum_{t=k+1}^{n} c_{it} q_t;$$
$$q^*_j = \sum_{t=1}^{k-1} c_{jt} q_t + d c_{jk} + (q_k - d) c_{jk} + \sum_{t=k+1}^{n} c_{jt} q_t.$$

Hence we obtain

$$q^*_i \not> c \left(\sum_{t=1}^{n} c_{it} \right) - (c - q_k) c_{ik} = c - (c - q_k) c_{ik}; \tag{5}$$

$$q^*_j \not< d \left(\sum_{t=1}^{n} c_{jt} \right) + (q_k - d) c_{jk} = d + (q_k - d) c_{jk}. \tag{6}$$

Using (4), (5) and (6), we obtain

$$q^*_i - q^*_j \not> c - d - (c - q_k)\delta_k - (q_k - d)\delta_k = (1 - \delta_k)\varkappa(\mathbf{Q}).$$

By Property 1 in Chap. I, §2,

$$\varkappa(\mathbf{CQ}) = \max_{i,j} (q^*_i - q^*_j).$$

But since

$$q^*_i - q^*_j \not> (1 - \delta_k)\varkappa(\mathbf{Q})$$

for any pair i, j, $1 \leqslant i, j \leqslant n$, it follows from Property 2 in Chap. I, §2, that

$$\varkappa(\mathbf{CQ}) \not> (1 - \delta_k)\varkappa(\mathbf{Q}),$$

as required.

Now let M_m^n be a quasidefinite system of stochastic $n \times n$ matrices with quasidefinite index $\boldsymbol{\mu}$. Every stochastic matrix $\mathbf{D}$ defined by

$$\begin{cases} \mathbf{D}=\mathbf{M}_x; \\ l(x)=\mu, \end{cases} \tag{7}$$

contains a positive column.

Let $[D_s]$ be the system of all matrices $\mathbf{D}$ defined as in (7). Since $M_m{}^n$ contains only finitely many different matrices, the same holds for the system $[D_s]$. Define a number s_c by

$$s_c = \max_{\mathbf{D}\in[D_s]} \varkappa(\mathbf{D}), \tag{8}$$

Suppose that $[D_s]$ contains exactly n^* distinct matrices, numbered in succession from 1 to n^*. Then we can effectively construct a matrix $\mathbf{D}^*$ with n^* columns $\mathbf{B}_i$, $i=1, 2, \ldots, n^*$, and an arithmetical function $\varphi(i)$, with the following properties: $\mathbf{B}_i$ is a positive column of $\mathbf{D}^*$ and it is the $\varphi(i)$-th column of the matrix $\dot{\mathbf{D}}_i\in[D_s]$. Letting d^*_{ij}, $i=1, 2, \ldots, n$, $j=1, 2, \ldots, n^*$, denote the entries of the matrix $\mathbf{D}^*$, we set

$$\delta_c = \min_{i,j} d^*_{ij}. \tag{9}$$ *

Theorem 2. Let $M_m{}^n$ be a quasidefinite system of stochastic $n\times n$ matrices with quasidefinite index μ, and $\mathbf{C}$ the product of r matrices in $M_m{}^n$. Then, if $r=\mu h$, where h is an arbitrary nonzero natural number,

$$\varkappa(\mathbf{C}) \not> (1-\delta_c)^{h-1}s_c. \tag{10}$$

Proof. By assumption, $\mathbf{C}$ is the product of h matrices in $[D_s]$:

$$\mathbf{C}=\mathbf{D}_{i_1}\mathbf{D}_{i_2}\ldots\mathbf{D}_{i_h}; \quad \mathbf{D}_{i_j}\in[D_s], \quad j=1, 2, \ldots, h. \tag{11}$$

Thus formula (10) is trivial for $h=1$. Supposing it true for some $h\geqslant 1$, let us prove it for $h+1$. By assumption,

$$\mathbf{C}=\mathbf{D}_{i_1}\mathbf{D}_{i_2}\ldots\mathbf{D}_{i_{h+1}}=\mathbf{D}_{i_1}\mathbf{C}',$$

where

$$\mathbf{C}'=\mathbf{D}_{i_2}\ldots\mathbf{D}_{i_{h+1}}.$$

By the inductive hypothesis,

$$\varkappa(C') \not> (1-\delta_c)^{h-1}s_c.$$

By the definition (9) of δ_c, we see that the matrix $\mathbf{D}_{i_1}$ contains a positive column $\mathbf{B}_k$ with entries $d_{1k}, d_{2k}, \ldots, d_{nk}$ such that

* Unlike s_c, the number δ_c is not uniquely defined. This is because there are quasidefinite systems for which the system $[D_s]$ contains matrices $\mathbf{D}$ with several positive columns.

$$\min_{i=1,2,\ldots,n} d_{ik}=\delta_k \not< \delta_c. \tag{12}$$

By Lemmas 1 and 2, it follows from (11) and (12) that

$$\varkappa(\mathbf{C}) \not> (1-\delta_c)\varkappa(\mathbf{C}').$$

Hence, by the inductive hypothesis,

$$\varkappa(\mathbf{C}) \not> (1-\delta_c)^h s_c.$$

Thus formula (10) holds for any natural number $h>0$.

The following theorem enables one to prove a converse of Theorem 2 and also provides a certain test for matrices to have positive columns.

Theorem 3. A stochastic $n\times n$ matrix $\mathbf{M}=(a_{ij})$ has a positive column if

$$\varkappa(\mathbf{M})<\frac{1}{n}. \tag{13}$$

Proof. By (13), there is a natural number m such that

$$\frac{1}{n}-\varkappa(\mathbf{M})>\frac{1}{m}. \tag{14}$$

Let $a_{1i}(\nu)$ denote the ν-th term in a constructive sequence of rational numbers defining a_{1i}.

Let ν^* be a natural number such that, for any $i=1, 2, \ldots, n$ and any $k\geqslant 0$,

$$|a_{1i}(\nu^*)-a_{1i}(\nu^*+k)|<\frac{1}{4m}. \tag{15}$$

By (15), there exists a value i, say i^*, for which

$$a_{1i^*}(\nu^*)>\frac{1}{n}-\frac{1}{2m}.$$

Indeed, if $a_{1i}(\nu^*)\leqslant\frac{1}{n}-\frac{1}{2m}$ for all $i=1, 2, \ldots, n$, then by (15), for all $i=1, 2, \ldots, n$ and any $k\geqslant 0$

$$a_{1i}(\nu^*+k)<\frac{1}{n}-\frac{1}{2m}+\frac{1}{4m}=\frac{1}{n}-\frac{1}{4m}.$$

Consequently,

$$\sum_{i=1}^{n} a_{1i} \not> 1-\frac{n}{4m}<1.$$

But this contradicts our assumptions, proving the existence of i^*. Then, by (15), we obtain the following bound for a_{1i^*}:

$$a_{1i^*} \nless \frac{1}{n} - \frac{3}{4m}.$$

Hence, by (14),

$$a_{1i^*} - \varkappa(\mathbf{M}) > \frac{1}{4m}. \tag{16}$$

Now $\varkappa(\mathbf{M}) \nless a_{1i^*} - a_{ji^*}$ for any $j = 1, 2, \ldots, n$, and so, by (16), $a_{ji^*} > \frac{1}{4m}$.

Thus, all the entries in the i^*-th column of the matrix $\mathbf{M}$ are strictly positive.

§4. QUASIDEFINITE FINITE STOCHASTIC AUTOMATA

In a now classical paper, Bukharaev (1965) considered the question of what events may be represented in an FSA. His solution is unfortunately unacceptable to constructive mathematics, so that a constructive solution must adopt a wholly new approach. However, we shall confine ourselves to analysis of a more restricted problem: what are the events that an FSA with quasidefinite system M_m^n represents with isolated cut-point λ? The classical solution of this problem may be found in Paz (1966a). We shall show that the constructive approach yields (formally) the same solution.

Concerning quasidefinite FSA (i. e., FSA for which M_m^n is quasidefinite) with a nonisolated cut-point λ, little is known. All that is known to date is that they may represent both regular and nonregular events.*

Let

$$\mathcal{A} = [\Sigma, S, M_2^2, \pi_0, \alpha]$$

be an FSA with

$$\mathbf{M}_{\sigma_1} = \begin{pmatrix} 1 & 0 \\ \frac{1}{2} & \frac{1}{2} \end{pmatrix}; \qquad \mathbf{M}_{\sigma_2} = \begin{pmatrix} \frac{1}{2} & \frac{1}{2} \\ 0 & 1 \end{pmatrix};$$

$$\pi_0 = (1, 0); \qquad \alpha = \begin{pmatrix} 0 \\ 1 \end{pmatrix}.$$

It is easy to see that this FSA represents a regular event with cut-point $\lambda = \frac{1}{2}$, i. e., the event $T\left(\mathcal{A}, \frac{1}{2}\right)$ is regular.

Indeed, the event $T\left(\mathcal{A}, \frac{1}{2}\right)$ is representable in the FDA

$$\mathcal{A}' = [\Sigma, S', M_2^3, \pi'_0, \alpha'],$$

* An event is said to be regular if it is representable in an FDA, i.e., an FSA in which the entries of $\mathbf{M}_{\sigma_i}$ and π_0 are integers.

where

$$\mathbf{M}_{\sigma_1} = \begin{pmatrix} 1 & 0 & 0 \\ 0 & 1 & 0 \\ 0 & 1 & 0 \end{pmatrix}; \qquad \mathbf{M}_{\sigma_2} = \begin{pmatrix} 0 & 1 & 0 \\ 0 & 0 & 1 \\ 0 & 0 & 1 \end{pmatrix};$$

$$\pi'_0 = (1, 0, 0); \qquad \alpha' = \begin{pmatrix} 0 \\ 0 \\ 1 \end{pmatrix}.$$

On the other hand, the event representable by $\mathcal{A}$ with cut-point $\lambda = \sqrt{2} - 1$ is not regular — there cannot exist an FDA representing the event $T(\mathcal{A}, \sqrt{2} - 1)$. A proof that neither of these cut-points are isolated may be found in Rabin (1963).

Definition. Let $\Sigma = (\sigma_1, \sigma_2, \ldots, \sigma_m)$ be an alphabet and U a set of words over Σ. We shall call U a *definite event** if there is a natural number h with the property: any word $z = yx$, where x is of length $l(x) = h$, is in U if and only if x is in U.

Note that in form the constructive definition of a definite event is exactly the same as the classical one, but in order to prove that an event U is definite in the constructive case one must provide an effective procedure producing the natural number h. In the absence of such a procedure, the proof is unacceptable.

Theorem 4. *Any event representable in an FSA with quasidefinite $M_m{}^n$ with an isolated cut-point is definite; in other words, quasidefinite FSA with an isolated cut-point λ represent only definite events.*

Proof. Let $\mathbf{C} = (c_{ij})$ be a matrix whose entries are CRNs. Let $\omega(\mathbf{C})$ denote the number $\max_{i,j} |c_{ij}|$.

The following three properties of stochastic $n \times n$ matrices $\mathbf{A} = (a_{ij})$, $\mathbf{B} = (b_{ij})$, $\mathbf{C} = (c_{ij})$ are obvious:

$$\omega(\mathbf{A} - \mathbf{A}) = 0; \tag{17}$$

$$\omega(\mathbf{A} - \mathbf{B}) = \omega(\mathbf{B} - \mathbf{A}); \tag{18}$$

$$\omega(\mathbf{A} - \mathbf{C}) \not> \omega(\mathbf{A} - \mathbf{B}) + \omega(\mathbf{B} - \mathbf{C}). \tag{19}$$

Somewhat less trivial is the property

$$\omega(\mathbf{BA} - \mathbf{CA}) \not> \varkappa(\mathbf{A}). \tag{20}$$

To prove (20), we let d_{ij} and g_{ij} denote the entries of $\mathbf{BA}$ and $\mathbf{CA}$, respectively. We claim that for any $j = 1, 2, \ldots, n$,

$$\min_i a_{ij} \not> d_{ij}, \quad g_{ij} \not> \max_i a_{ij}. \tag{21}$$

* The above event $T\left(\mathcal{A}, \frac{1}{2}\right)$ is an example of a regular event which is not definite.

Indeed, since

$$d_{ij}=\sum_{t=1}^{n} b_{it}a_{tj},$$

it follows that

$$(\min_t a_{tj})\left(\sum_{t=1}^{n} b_{it}\right)\ngtr d_{ij}\ngtr(\max_t a_{tj})\left(\sum_{t=1}^{n} b_{it}\right).$$

Hence, since $\mathbf{B}$ is stochastic, we have

$$\min_t a_{tj}\ngtr d_{ij}\ngtr\max_t a_{tj}.$$

Similar arguments hold for g_{ij}. Consequently, formula (21) is valid for any i, j, $1\leqslant i$, $j\leqslant n$. We now infer from (21) that

$$|d_{ij}-g_{ij}|\ngtr\varkappa(\mathbf{A}). \tag{22}$$

Hence, by Property 2 in Chap. I, §2,

$$\max_{i,j}|d_{ij}-g_{ij}|\ngtr\varkappa(\mathbf{A}).$$

We now proceed directly to prove the theorem itself. By assumption, the matrices $\mathbf{M}_{\sigma_j}$ form a quasidefinite system of stochastic $n\times n$ matrices. Let μ be a quasidefinite index of the system $M_m{}^n$. Then, by Theorem 3, for any word of length $l(x)=h\mu$ (where h is a natural number) over Σ,

$$\varkappa(\mathbf{M}_x)\ngtr(1-\delta_c)^{h-1}s_c.$$

Hence, by (20), for any word y over Σ,

$$\omega(\mathbf{M}_{yx}-\mathbf{M}_x)\ngtr(1-\delta_c)^{h-1}s_c. \tag{23}$$

Suppose now that the FSA $\mathcal{A}$ has exactly l final states $s_{i_1}, s_{i_2}, \ldots, s_{i_l}$, i. e., $\alpha_{i_1}=\alpha_{i_2}=\ldots=\alpha_{i_l}=1$. Then, provided the word x has length $h\mu$, it follows from (23) that for any y

$$|p_{\mathcal{A}}(yx)-p_{\mathcal{A}}(x)|\ngtr(1-\delta_c)^{h-1}s_c l. \tag{24}$$

Let λ, $0<\lambda<1$, be a number such that for any word z over Σ and some rational number $r>0$

$$|p_{\mathcal{A}}(z)-\lambda|>r. \tag{25}$$

We claim that the event $T(\mathcal{A},\lambda)$ is definite. In other words, we must (effectively!) find a natural number h^* such that

$$(1-\delta_c)^{h^*-1}s_c l<r, \tag{24'}$$

and show that any word $z=yx$, $l(x)=h^*\mu$, lies in $T(\mathcal{A}, \lambda)$ if and only if $x\in T(\mathcal{A}, \lambda)$. Indeed, if $x\in T(\mathcal{A}, \lambda)$ then, by (24) and (24'),

$$\lambda<p_{\mathcal{A}}(x)-r;$$

$$p_{\mathcal{A}}(x)-p_{\mathcal{A}}(yx)<r.$$

Consequently,

$$\lambda<p_{\mathcal{A}}(x)-r<p_{\mathcal{A}}(yx). \tag{26}$$

Thus, if x is a word of length $l(x)=h^*\mu$ in $T(\mathcal{A}, \lambda)$, then any word $z=yx$ is also in $T(\mathcal{A}, \lambda)$.

Suppose now that $z=yx$ is in $T(\mathcal{A},\lambda)$, where x is of length $h^*\mu$; then, by (25), either $p_{\mathcal{A}}(x)+r<\lambda$ or $\lambda<p_{\mathcal{A}}(x)-r$.

If $p_{\mathcal{A}}(x)+r<\lambda$, then $p_{\mathcal{A}}(yx)<\lambda$, for by (24) and (24'),

$$p_{\mathcal{A}}(yx)<p_{\mathcal{A}}(x)+r.$$

This contradiction shows that the second alternative must hold. Consequently, it follows from $z\in T(\mathcal{A}, \lambda)$ that $p_{\mathcal{A}}(x)>\lambda$. But this means that $x\in T(\mathcal{A}, \lambda)$, and the proof is complete.

Theorem 4 shows that FSA which represent regular indefinite events with an isolated cut-point are in a sense "strongly deterministic," i. e., "many" of their state transitions are forbidden. This conclusion is also of practical significance: structural synthesis of an FSA representing regular indefinite events with an isolated cut-point is possible only if the basis for the synthesis contains a certain subsystem of strongly deterministic function elements, or elements whose performance can be made deterministic under certain working conditions.

Theorem 4 brings us back to the question of whether an FSA saves states in comparison with an FDA representing the same event (when the former has an isolated cut-point).

The following is a direct corollary of Theorem 3 (§ 3) and formula (20): The entries of products of stochastic matrices comprising a quasidefinite system may be computed with given precision $\varepsilon=2^{-k}$ by a computer with limited memory.*

One must bear in mind, however, an important feature of the phrase "may be computed . . . by a computer with limited memory." In so saying, we are ignoring a significant component of the problem as it arises in practice, for no attention is given to determining the requisite memory capacity of the computer. Without a solution to this problem the whole computational problem in fact remains unsolved. The main import of the above phrase is thus that in the final analysis any computational problem of the type considered demands use of a computer whose memory cannot be restricted in advance.

* The system of stochastic matrices to which Rabin (1963) applies his approximation scheme is quasidefinite.

§5. STABILITY OF FINITE STOCHASTIC AUTOMATA

From our constructivist standpoint, the formulation of the stability problem by Rabin (1963) is far from satisfactory. Nevertheless, the underlying ideas of his classical definitions of stability lead the way to formulation of the problem in an acceptable constructive form. Moreover, the constructive definition partly retains the form of the classical one, except that the latter is of course understood in terms of constructive conceptions.

Consider an FSA

$$\mathcal{A}=[\Sigma, S, \mathbf{M}_m{}^n, \pi_0, a]$$

and a given rational number $\varepsilon>0$. We let $\{\mathfrak{A}_\varepsilon\}$ denote the class of FSA

$$\mathcal{A}'=[\Sigma, S, M'_m{}^n, \pi_0, a]$$

satisfying the conditions:

(1) for any letter σ of the alphabet Σ,

$$\omega(\mathbf{M}_\sigma-\mathbf{M}'_\sigma) \not> \varepsilon;$$

(2) for any letter σ of Σ and any entry $a_{ij}(\sigma)$ of the matrix $\mathbf{M}_\sigma$, if $a_{ij}(\sigma)$ vanishes, then so does the entry $a'_{ij}(\sigma)$ of $\mathbf{M}'_\sigma$.

Definition 6. Let $\mathcal{A}$ be an FSA representing an event U with isolated cut-point λ, i. e., $U=T(\mathcal{A}, \lambda)$. We shall say that $\mathcal{A}$ is stable relative to λ if and only if there exist rational numbers $\varepsilon>0$ and $\delta>0$ such that any FSA $\mathcal{A}'\in\{\mathfrak{A}_\varepsilon\}$ satisfies the condition $\forall x(|p_{\mathcal{A}'}(x)-\lambda| \not< \delta)$.

If we omit condition (2) in the definition of the system $\{\mathfrak{A}_\varepsilon\}$, the concept of stability as defined in Definition 6 becomes stronger. We shall therefore refer to this variant of stability as strong stability relative to λ.

It is not hard to construct an FSA which is stable relative to any isolated cut-point λ but not strongly stable relative to the same values λ. An example is the FSA representing the event $T\left(\mathcal{A}, \frac{1}{2}\right)$ constructed at the beginning of §4. It may nevertheless be shown (see Theorem 6) that there is a whole class of FSA which are strongly stable relative to any isolated cut-point λ. In the classical context, this problem was solved by Paz (1966a).

Theorem 5 below reveals a completely different relation between the conditions defining stability. It shows that the assumption $U=T(\mathcal{A}', \lambda)$ is superfluous: it follows from the fact that λ is isolated for $\mathcal{A}'\in\{\mathfrak{A}_\varepsilon\}$.

Theorem 5. Let $\mathcal{A}$ be an FSA, λ a cut-point and ε a positive rational number. If λ is an isolated cut-point for each FSA $\mathcal{A}'\in\{\mathfrak{A}_\varepsilon\}$, then

$$T(\mathcal{A}', \lambda)\approx T(\mathcal{A}, \lambda).$$

Proof. Let

$$\mathcal{A}=[\Sigma, S, M_m{}^n, \pi_0, a]$$

be an FSA representing the event U with isolated cut-point λ, and $\varepsilon > 0$ a rational number such that the class $\{\mathfrak{A}_\varepsilon\}$ contains only FSA for which λ is an isolated cut-point in representations of events $U' \approx T(\mathcal{A}, \lambda)$. We claim that under these assumptions $U' \approx U$.

Let $\mathcal{A}_1$ and $\mathcal{A}_2$ be FSA of the class $\{\mathfrak{A}_\varepsilon\}$. Let $\mathbf{B}_{\sigma_i}$ and $\mathbf{C}_{\sigma_i}$ be the matrices defined by

$$\mathbf{B}_{\sigma_i} = \mathbf{M}_{\sigma_i} - (\mathbf{M}_1)_{\sigma_i}; \tag{27}$$

$$\mathbf{C}_{\sigma_i} = \mathbf{M}_{\sigma_i} - (\mathbf{M}_2)_{\sigma_i}. \tag{28}$$

We shall show that the FSA

$$\tilde{\mathcal{A}} = [\Sigma, S, M_m{}^n, \pi_0, \alpha]$$

with transition matrices $\tilde{\mathbf{M}}_{\sigma_i}$ defined by

$$\tilde{\mathbf{M}}_{\sigma_i} = \mathbf{M}_{\sigma_i} - \frac{1}{2}(\mathbf{B}_{\sigma_i} + \mathbf{C}_{\sigma_i}), \tag{29}$$

is also in $\{\mathfrak{A}_\varepsilon\}$.

Indeed, since

$$|b_{kj}(\sigma_i)| \ngtr \varepsilon; \qquad |c_{kj}(\sigma_i)| \ngtr \varepsilon$$

it follows at once that

$$\frac{1}{2}|b_{kj}(\sigma_i) + c_{kj}(\sigma_i)| \ngtr \varepsilon.$$

Hence we conclude that the matrices $\tilde{\mathbf{M}}_{\sigma_i}$ satisfy condition (1).

Since

$$b_{kj}(\sigma_i) = c_{ki}(\sigma_i) = 0$$

provided $a_{kj}(\sigma_i) = 0$, we have

$$\tilde{a}_{kj}(\sigma_i) = a_{kj}(\sigma_i).$$

Consequently, the matrices $\tilde{\mathbf{M}}_{\sigma_i}$ also satisfy condition (2). The fact that the matrices $\tilde{\mathbf{M}}_{\sigma_i}$ are stochastic is a direct consequence of the following easily verified relations:

$$\sum_{j=1}^{n} b_{kj}(\sigma_i) = \sum_{j=1}^{n} c_{kj}(\sigma_i) = 0; \tag{30}$$

$$0 \ngtr a_{kj}(\sigma_i) - b_{kj}(\sigma_i) \ngtr 1; \tag{31}$$

$$0 \ngtr a_{kj}(\sigma_i) - c_{kj}(\sigma_i) \ngtr 1. \tag{32}$$

Indeed, by (30),

$$\sum_{j=1}^{n} \tilde{a}_{kj}(\sigma_i) = \sum_{j=1}^{n} a_{kj}(\sigma_i) = 1,$$

and by (31) and (32),

$$0 \ngtr \tilde{a}_{kj}(\sigma_i) \ngtr 1.$$

Thus the automaton $\tilde{\mathcal{A}}$ is indeed in the class $\{\mathfrak{A}_\varepsilon\}$.

The properties of matrix multiplication show that for any word x over Σ we can determine a polynomial $P^x(y_{11}(\sigma_1), \ldots, y_{nn}(\sigma_m))$, the number of whose arguments $y_{kj}(\sigma_i)$ is $\widehat{n} = mn^2$ (m is the number of input states, n the number of internal states of $\mathcal{A}$), such that for any FSA $\mathcal{A}^* \in \{\mathfrak{A}_\varepsilon\}$

$$p_{\mathcal{A}^*}(x) = P^x((a_{11}(\sigma_1) - a^*_{11}(\sigma_1)), \ldots, (a_{nn}(\sigma_m) - a^*_{nn}(\sigma_m))). \quad (33)$$

Clearly, for $\mathcal{A}$ itself we obtain

$$p_{\mathcal{A}}(x) = P^x(0, \ldots, 0).$$

Note that in general the polynomial $P_x(y_{11}(\sigma_1), \ldots, y_{nn}(\sigma_m))$ does not have rational coefficients; but if the matrices $\mathbf{M}_{\sigma_i}$ in the system $M_m{}^n$ for $\mathcal{A}$ have only rational entries, all the coefficients of these polynomials will be rational.*

Suppose now that there exist a word x over Σ and an FSA $\mathcal{A}' \in \{\mathfrak{A}_\varepsilon\}$ such that either

$$p_{\mathcal{A}}(x) > \lambda; \qquad p_{\mathcal{A}'}(x) \ngtr \lambda, \quad (34)$$

or

$$p_{\mathcal{A}}(x) < \lambda; \qquad p_{\mathcal{A}'}(x) \nless \lambda. \quad (35)$$

By the assumptions of the theorem, there exists a rational number $\delta' > 0$ such that either

$$p_{\mathcal{A}}(x) > \lambda; \qquad p_{\mathcal{A}'}(x) < \lambda - \delta', \quad (34')$$

or

$$p_{\mathcal{A}}(x) < \lambda; \qquad p_{\mathcal{A}'}(x) > \lambda + \delta'. \quad (35')$$

We now show that the FSA $\mathcal{A}'$ satisfying conditions (34) and (35) may be so chosen that all entries of the matrices $\mathbf{M}_{\sigma_i} - \mathbf{M}'_{\sigma_i}$ are rational numbers.

Indeed, by (33),

$$p_{\mathcal{A}'}(x) = P^x((a_{11}(\sigma_1) - a'_{11}(\sigma_1)), \ldots, (a_{nn}(\sigma_m) - a'_{nn}(\sigma_m))).$$

* We are assuming that the polynomial does not contain terms which differ only as to their coefficients.

Consequently, by Tseitin's theorem on continuity of constructive functions (Tseitin, 1962a), we can find a natural number l such that

$$|p_{\mathcal{A}'}(x) - P^x(r_{11}(\sigma_1), \ldots, r_{nn}(\sigma_m))| < \frac{\delta'}{2}, \tag{36}$$

provided the rational numbers $r_{kj}(\sigma_i)$ satisfy the inequality

$$|a_{kj}(\sigma_i) - a'_{kj}(\sigma_i) - r_{kj}(\sigma_i)| < \frac{1}{l}. \tag{37}$$

We denote the numbers $(a_{kj}(\sigma_i) - a'_{kj}(\sigma_i))$ by $b'_{kj}(\sigma_i)$ and for each of them construct a rational number $r_{kj}(\sigma_i)$ with the property

$$|b'_{kj}(\sigma_i) - r_{kj}(\sigma_i)| < \frac{1}{2\nu^2}, \tag{38}$$

where

$$\nu = \max(2l, n). \tag{39}$$

Next, for each number $r_{kj}(\sigma_i)$ satisfying (38) we construct a rational number $r'_{kj}(\sigma_i)$ as follows: $r'_{kj}(\sigma_i) = 0$ if

$$|r_{kj}(\sigma_i)| \leqslant \frac{1}{2\nu^2};$$

otherwise

$$r'_{kj}(\sigma_i) = r_{kj}(\sigma_i) - \operatorname{sign}(r_{kj}(\sigma_i)) \frac{1}{2\nu^2}.$$

The numbers $r'_{kj}(\sigma_i)$ are easily seen to satisfy the inequality

$$|b'_{kj}(\sigma_i) - r'_{kj}(\sigma_i)| < \frac{1}{\nu^2}. \tag{40}$$

Consequently, for any fixed numbers k and i, $1 \leqslant k \leqslant n$, $1 \leqslant i \leqslant m$,

$$-\frac{1}{\nu} < \sum_{j=1}^{n} r'_{kj}(\sigma_i) < \frac{1}{\nu}. \tag{41}$$

Now let η_{ki} denote the numbers defined by

$$\eta_{ki} = \sum_{j=1}^{n} r'_{kj}(\sigma_i).$$

We shall show that if all these numbers η_{ki} vanish, the required FSA is

$$\mathcal{A}'' = [\Sigma, S, M''^{n}_{m}, \pi_0, \alpha],$$

where the entries of the matrices $\mathbf{M}''_{\sigma_i}$ are defined by

$$a''_{kj}(\sigma_i) = a_{kj}(\sigma_i) - r'_{kj}(\sigma_i).$$

Indeed, if some entry $a_{kj}(\sigma_i)$ is zero, the same is true of $b'_{kj}(\sigma_i)$. Hence, by (38),

$$|r_{kj}(\sigma_i)| = |r_{kj}(\sigma_i) - b'_{kj}(\sigma_i)| < \frac{1}{2\nu^2}.$$

Consequently, it follows from the definition of $r'_{kj}(\sigma_i)$ that $r'_{kj}(\sigma_i) = 0$. This clearly demonstrates that the FSA $\mathcal{A}''$ satisfies condition (2).

We next show that the numbers $r'_{kj}(\sigma_i)$ all satisfy the inequality

$$|r'_{kj}(\sigma_i)| \not> |b'_{kj}(\sigma_i)|. \tag{42}$$

Indeed, if $r'_{kj}(\sigma_i) = 0$ this is trivial, so we assume that $r'_{kj}(\sigma_i) \neq 0$. By formula (38),

$$|r_{kj}(\sigma_i)| < |b'_{kj}(\sigma_i)| + \frac{1}{2\nu^2},$$

whence, by the definition of $r'_{kj}(\sigma_i)$, we obtain

$$|r'_{kj}(\sigma_i)| = |r_{kj}(\sigma_i)| - \frac{1}{2\nu^2} < |b'_{kj}(\sigma_i)|.$$

Thus the matrices $\mathbf{M}''_{\sigma_i}$ satisfy condition (1).

To complete the proof, it remains only to show that the matrices $\mathbf{M}''_{\sigma_i}$ are stochastic and that $p_{\mathcal{A}''}(x)$ satisfy one of conditions (34), (35). We begin with the proof that the matrices $\mathbf{M}''_{\sigma_i}$ are stochastic.

Let $a''_{kj}(\sigma_i) < 0$. Then, using the fact that the entries $a_{kj}(\sigma_i)$ are nonnegative and condition (42), we obtain

$$a_{kj}(\sigma_i) < r'_{kj}(\sigma_i) \not> |b'_{kj}(\sigma_i)|.$$

Hence, if $r'_{kj}(\sigma_i) = 0$ then $a_{kj}(\sigma_i) < 0$; if $r'_{kj}(\sigma_i) > 0$ then, by the definition of $r'_{kj}(\sigma_i)$ and formula (42),

$$a_{kj}(\sigma_i) - b'_{kj}(\sigma_i) < 0.$$

Consequently, there cannot be any entries with $a''_{kj}(\sigma_i) < 0$. Thus, for all k, j, i,

$$0 \not> a''_{kj}(\sigma_i). \tag{43}$$

Suppose now that there exists an entry with $a''_{kj}(\sigma_i) > 1$. Then clearly $r'_{kj}(\sigma_i) < 0$. But then, by the definition of $r'_{kj}(\sigma_i)$ and (42),

$$b'_{kj}(\sigma_i) < r'_{kj}(\sigma_i).$$

Consequently,

$$a_{kj}(\sigma_i) - b'_{kj}(\sigma_i) \not< a''_{kj}(\sigma_i) > 1,$$

which is impossible. Thus, for all k, j, i,

$$a''_{kj}(\sigma_i) \ngtr 1. \tag{44}$$

Thus, by our assumption on the numbers η_{ki} and by (43), (44), we have proved that the matrices $\mathbf{M}''_{\sigma_i}$ are indeed stochastic.

To fix ideas, let us assume that the FSA $\mathcal{A}'$ satisfies conditions (34) and (34'). Then

$$P^x(b'_{11}(\sigma_1), \ldots, b'_{nn}(\sigma_m)) < \lambda - \delta'. \tag{45}$$

In view of (36), (37), (39), (40), we obtain

$$|P^x(b'_{11}(\sigma_1), \ \ldots, \ b'_{nn}(\sigma_m)) - P^x(r'_{11}(\sigma_1), \ \ldots, \ r'_{nn}(\sigma_m))| < \frac{\delta'}{2}.$$

Hence, by (45), (33),

$$p_{\mathcal{A}''}(x) = P^x(r'_{11}(\sigma_1), \ \ldots, \ r'_{nn}(\sigma_m)) < \lambda - \frac{\delta'}{2}.$$

We omit the arguments for the case that condition (35) holds — they are entirely analogous.

However, it is quite possible that $\eta_{ki} \neq 0$ for some k, i. Then either $\eta_{ki} > 0$ or $\eta_{ki} < 0$.

Suppose the latter holds. Then the sum $\tilde{\eta}_{ki}$ of all negative terms $r'_{kj}(\sigma_i)$, in the sum for η_{ki} satisfies the inequality $|\tilde{\eta}_{ki}| \geqslant |\eta_{ki}|$.

Consequently, by (41), subtracting at most min $(|r'_{ki}(\sigma_i)|, |\eta_{ki}|)$ from the absolute values of the negative elements $r'_{kj}(\sigma_i)$ we replace the numbers $r'_{k1}(\sigma_i), \ldots, r'_{kn}(\sigma_i)$ by a new system of rational numbers $\varrho_{k1}(\sigma_i), \ldots, \varrho_{kn}(\sigma_i)$, such that

$$\sum_{j=1}^{n} \varrho_{kj}(\sigma_i) = 0.$$

This leaves all positive and vanishing terms $r'_{kj}(\sigma_i)$ unchanged:

$$\varrho_{kj}(\sigma_i) = r'_{kj}(\sigma_i),$$

provided $r'_{kj}(\sigma_i) \geqslant 0$.

If $\eta_{ki} > 0$, we proceed in analogous fashion and again obtain numbers $\varrho_{kj}(\sigma_i)$ such that

$$\sum_{j=1}^{n} \varrho_{kj}(\sigma_i) = 0.$$

Now it is the nonpositive terms $r'_{kj}(\sigma_i)$ that remain unchanged, i. e., $\varrho_{kj}(\sigma_i) = r'_{kj}(\sigma_i)$ if $r'_{kj}(\sigma_i) \leqslant 0$.

The following properties of the numbers $\varrho_{kj}(\sigma_i)$ are easily proved:

$$|\varrho_{kj}(\sigma_i)| \ngtr |r'_{kj}(\sigma_i)| \ngtr |b'_{kj}(\sigma_i)|; \tag{46}$$

$$|b'_{kj}(\sigma_i) - \varrho_{kj}(\sigma_i)| < \frac{1}{l}. \qquad (47)$$

Finally, defining the numbers $\varrho_{kj}(\sigma_i)$ by

$$\varrho_{kj}(\sigma_i) = r'_{kj}(\sigma_i)$$

for values k, i such that $\eta_{ki}=0$, we complete construction of a system of rational numbers $\varrho_{kj}(\sigma_i)$ satisfying (46) and (47). Now, replacing the numbers $r'_{kj}(\sigma_i)$ in the definition of $\mathcal{A}''$ by $\varrho_{kj}(\sigma_i)$, we again obtain an FSA. It is readily shown that this automaton possesses the same basic properties as $\mathcal{A}''$, and so the construction is complete.

We may thus assume that the original automaton $\mathcal{A}'$ was so chosen that the matrices $\mathbf{M}_{\sigma_i} - \mathbf{M}'_{\sigma_i}$ contain only rational entries, i. e., all the numbers $b'_{kj}(\sigma_i)$ are rational. Let $\widetilde{P}^x(y_{11}(\sigma_1), \ldots, y_{nn}(\sigma_m))$ denote the polynomial

$$P^x(y_{11}(\sigma_1), \ldots, y_{nn}(\sigma_m)) - \lambda.$$

By our assumptions concerning $\mathcal{A}'$, the polynomial $P^x(y_{11}(\sigma_1), \ldots, y_{nn}(\sigma_m))$ assumes values with opposite signs at the points p_0 and p_1 with coordinates $y_{kj}(\sigma_i)=0$ and $y'_{kj}(\sigma_i)=b'_{kj}(\sigma_i)$, respectively. Let us view these as the first two terms in a sequence of points $p_0, p_1, \ldots, p_{2t}, p_{2t+1}, \ldots$ in euclidean mn^2-space. The remaining terms of the sequence are defined inductively: if the coordinates of p_{2t} are $u_{kj}(\sigma_i)$ and those of p_{2t+1} are $v_{kj}(\sigma_i)$, then those of p_{2t+2} will be $\dfrac{u_{kj}(\sigma_i)+v_{kj}(\sigma_i)}{2}$. The coordinates of p_{2t+3} coincide with those of p_{2t+1} if the polynomial $\widetilde{P}^x(y_{11}(\sigma_1), \ldots, \ldots, y_{nn}(\sigma_m))$ assumes values with opposite signs at the point p_{2t+1} and p_{2t+2} Otherwise, the coordinates of p_{2t+3} coincide with those of p_{2t}.

The definition of the sequence $p_0, p_1, \ldots, p_{2t}, p_{2t+1}, \ldots$, is clearly effective. This follows directly from the fact that the coordinates of p_0, p_1 are rational, the properties of FSA, and the fact that λ is an isolated cut-point for any FSA of the class $\{\mathfrak{A}_\varepsilon\}$.

It is also easy to see that the sequence converges constructively to a point p with coordinates $w_{kj}(\sigma_i)$, say.

Let $\mathbf{W}_{\sigma_i}$ denote the $n\times n$ matrix whose k-th row is $w_{k1}(\sigma_i), w_{k2}(\sigma_i), \ldots, w_{kn}(\sigma_i)$ and $\mathbf{D}_{\sigma_i}$ the matrix defined by

$$\mathbf{D}_{\sigma_i} = \mathbf{M}_{\sigma_i} - \mathbf{W}_{\sigma_i}.$$

The properties of the FSA $\widetilde{\mathcal{A}}$ and the sequence p_0, p_1, $\ldots$, p_{2t}, p_{2t+1}, $\ldots$ show that $\mathbf{D}_{\sigma_i}$ is a stochastic matrix and satisfies conditions (1) and (2). Consequently, the FSA

$$\mathcal{A}_d = [\Sigma, S, M_m{}^n, \pi_0, \alpha]$$

with matrices $(\mathbf{M}_d)_{\sigma_i} = \mathbf{D}_{\sigma_i}$ is in class $\{\mathfrak{A}_\varepsilon\}$ and

$$p_{\mathcal{A}_d}(x) = P^x(w_{11}(\sigma_1), \ldots, w_{nn}(\sigma_m)).$$

But by the definition of the numbers $w_{kj}(\sigma_i)$, we have

$$\widetilde{P}^x(w_{11}(\sigma_1), \ldots, w_{nn}(\sigma_m)) = 0,$$

and so $p_{\mathcal{A}_d}(x) = \lambda$.

This of course contradicts the assumptions of the theorem, according to which λ is an isolated cut-point for any FSA in $\{\mathfrak{A}_\varepsilon\}$. Thus, for any word x over Σ, we have the following: if $p_{\mathcal{A}}(x) > \lambda$, then $p_{\mathcal{A}'}(x) \not> \lambda$ for every FSA $\mathcal{A}' \in \{\mathfrak{A}_\varepsilon\}$; if $p_{\mathcal{A}}(x) < \lambda$, then $p_{\mathcal{A}'}(x) \not< \lambda$ for every FSA $\mathcal{A}' \in \{\mathfrak{A}_\varepsilon\}$. Thus, since λ is an isolated cut-point for any FSA $\mathcal{A}' \in \{\mathfrak{A}_\varepsilon\}$, it follows that

$$T(\mathcal{A}', \lambda) \approx T(\mathcal{A}, \lambda).$$

This completes the proof of Theorem 5.

The reader will probably have noticed that the proof of Theorem 5 as presented above may be adapted without serious modifications to the case that the FSA in class $\{\mathfrak{A}_\varepsilon\}$ satisfy condition (1) but not necessarily condition (2).

§6. STABILITY OF FINITE STOCHASTIC AUTOMATA WITH QUASIDEFINITE SYSTEM OF TRANSITION MATRICES

In order to prove the main result of this section (Theorem 6), we shall first establish a lemma; in principle, the theorem may be proved without the lemma, but the latter is important for otherwise we would have to make do with less delicate estimates of various quantities used in the stability proof.

Lemma 2. Let $r > 0$ be a rational number, $a_1, a_2, \ldots, a_n$ and $b_1, b_2, \ldots, b_n$ CRNs satisfying the conditions:

(1) $\sum_{i=1}^{n} a_i = 0$;

(2) *for any i, $1 \leqslant i \leqslant n$,*

$$0 \not> b_i \not> 1; \quad |a_i| \not> r.$$

Then

$$\left| \sum_{i=1}^{n} a_i b_i \right| \not> r \cdot \frac{n}{2}. \tag{48}$$

Proof. Let c denote the number $\sum_{i=1}^{n} a_i b_i$ and suppose that $|c| > r \cdot \frac{n}{2}$. By the constructive interpretation of $>$, this means that we can construct a rational number $\varrho > 0$ with the property

$$|c| \not< r \cdot \frac{n}{2} + \varrho. \tag{49}$$

Let $a_i(m)$ denote the m-th term of a constructive sequence of rationals defining a_i; we can find a natural number m' such that for any i, $1 \leqslant i \leqslant n$, and any natural number ν,

$$|a_i(m') - a_i(m'+\nu)| < \frac{\varrho}{2n}. \tag{50}$$

Let I_1, I_2, I_3 denote the classes of indices i for which, respectively,

$$a_i(m') > \frac{\varrho}{2n}; \quad -\frac{\varrho}{2n} \leqslant a_i(m') \leqslant \frac{\varrho}{2n}; \quad a_i(m') < -\frac{\varrho}{2n}.$$

It is readily seen that $a_i > 0$ if $i \in I_1$; similarly, $a_i < 0$ if $i \in I_3$. By (50) and condition (2), for any $i \in I_2$,

$$|a_i b_i| \ngtr \frac{\varrho}{n}. \tag{51}$$

It now follows from (49), (51) and condition (2) that either

$$\sum_{i \in I_1} a_i b_i > r\frac{n}{2}, \tag{52}$$

or

$$\sum_{i \in I_3} |a_i b_i| > r\frac{n}{2}. \tag{53}$$

We shall show that (52) yields a contradiction, leaving discussion of (53) to the reader.

Suppose, then, that (52) holds. Then

$$\sum_{i \in I_1} a_i \nless \sum_{i \in I_1} a_i b_i > r\frac{n}{2}. \tag{54}$$

Let $\widehat{I_1}$ denote the number of indices in I_1; then, by condition (2) and (54) we have $\widehat{I_1} r > r \cdot \frac{n}{2}$. Consequently,

$$\widehat{I_1} \geqslant E\left(\frac{n}{2}\right) + 1.$$

Together with formula (54), this contradicts condition (1).

Thus the assumption that $|c| > r\frac{n}{2}$ yields a contradiction, proving (48).

Theorem 6. An FSA

$$\mathcal{A} = [\Sigma, S, M_m{}^n, \pi_0, \alpha]$$

with quasidefinite $M_m{}^n$ is strongly stable relative to every isolated cut-point λ.

Proof. Let λ, $0<\lambda<1$, be such that for any word x over Σ and some rational number $\delta>0$

$$|p_{\mathcal{A}}(x)-\lambda| \not< \delta. \tag{55}$$

We shall prove the existence of a rational number $\varepsilon>0$ such that λ is an isolated cut-point for every FSA

$$\mathcal{A}'=[\Sigma, S, M'^{n}_{m}, \pi_0, \alpha]$$

satisfying condition (1) of § 5, and

$$T(\mathcal{A}', \lambda) \approx T(\mathcal{A}, \lambda).$$

Let μ be a quasidefinite index of the system M_m^n. Then the system of matrices $[D_s]$ defined as before from M_m^n consists of all matrices $\mathbf{M}_x$ such that $l(x)=\mu$ (see § 2).

Define the number δ_c for $[D_s]$ and construct a rational number $r>0$ such that $r<\delta_c$. Let $d_c=\delta_c-r$, and denote by l the number of final states of $\mathcal{A}$ Find a natural number h such that

$$(1-d_c)^{h-1}<\frac{\delta}{6l}. \tag{56}$$

We claim that the required number ε may be taken equal to $\dfrac{\min\left(r, \dfrac{\delta}{6l}\right)}{(\mu h-1)\dfrac{n}{2}+1}$.

To prove this, we shall show that for any word x over Σ

$$\omega(\mathbf{M}_x-\mathbf{M}'_x) \not> \left((l(x)-1)\frac{n}{2}+1\right)\varepsilon, \tag{57}$$

provided only that, for any i, $1\leqslant i\leqslant n$,

$$\omega(\mathbf{M}_{\sigma_i}-\mathbf{M}'_{\sigma_i})<\varepsilon. \tag{58}$$

We prove (57) by induction on $l(x)$. If $l(x)=1$ the formula is trivially true. Supposing that it holds for any x of length $l(x)\leqslant k$, we consider an arbitrary word of the form $y=\sigma_i x$.

Define two new matrices $\mathbf{A}$ and $\mathbf{B}$ by

$$\mathbf{A}=\mathbf{M}'_{\sigma_i}-\mathbf{M}_{\sigma_i}; \tag{59}$$

$$\mathbf{B}=\mathbf{M}'_x-\mathbf{M}_x. \qquad (60)^*$$

By (59) and (60),

$$\mathbf{M}'_{\sigma_i}=\mathbf{M}_{\sigma_i}+\mathbf{A};$$

$$\mathbf{M}'_x=\mathbf{M}_x+\mathbf{B}.$$

* It is assumed here that i and x are fixed.

Hence $\mathbf{M}'_y$ is given by

$$\mathbf{M}'_y = (\mathbf{M}_{\sigma_i} + \mathbf{A})(\mathbf{M}_x + \mathbf{B}).$$

Consequently,

$$\omega(\mathbf{M}'_y - \mathbf{M}_y) = \omega(\mathbf{M}_{\sigma_i}\mathbf{B} + \mathbf{A}\mathbf{M}'_x).$$

By formula (19) (see §4), this now implies

$$\omega(\mathbf{M}'_y - \mathbf{M}_y) \ntriangleright \omega(\mathbf{M}_{\sigma_i}\mathbf{B}) + \omega(\mathbf{A}\mathbf{M}'_x). \tag{61}$$

By the inductive hypothesis,

$$\omega(\mathbf{B}) \ntriangleright \left((l(x)-1)\frac{n}{2}+1\right)\varepsilon,$$

whence, since $\mathbf{M}_{\sigma_i}$ is a stochastic matrix,

$$\omega(\mathbf{M}_{\sigma_i}\mathbf{B}) \ntriangleright \left((l(x)-1)\frac{n}{2}+1\right)\varepsilon. \tag{62}$$

It follows from (58) and (59) that for every entry a_{kj} in $\mathbf{A}$,

$$|a_{kj}| < \varepsilon. \tag{63}$$

On the other hand, since $\mathbf{M}_{\sigma_i}$ and $\mathbf{M}'_{\sigma_i}$ are stochastic matrices it follows that for any k, $0 \leqslant k \leqslant n$,

$$\sum_{j=1}^{n} a_{kj} = 0. \tag{64}$$

Using the stochastic properties of $\mathbf{M}'_x$ and (63) and (64), we infer from Lemma 2 that

$$\omega(\mathbf{A}\mathbf{M}'_x) \ntriangleright \varepsilon \cdot \frac{n}{2}. \tag{65}$$

Consequently, we see from (61), (62) and (65) that

$$\omega(\mathbf{M}'_y - \mathbf{M}_y) \ntriangleright \left((l(y)-1)\frac{n}{2}+1\right)\varepsilon.$$

Thus formula (57) is indeed valid for every word x over the alphabet Σ.

Suppose now that $\mathcal{A}'$ satisfies condition (1) (§5) with

$$\varepsilon = \frac{\min\left(r, \frac{\delta}{6l}\right)}{(\mu h - 1)\frac{n}{2} + 1}. \tag{66}$$

We claim that the matrices $\mathbf{M}'_{\sigma_i}$ form a quasidefinite system.

Indeed, by (57), (58) and (66), for any word x of length $l(x)=\mu$,

$$\omega(\mathbf{M}'_x-\mathbf{M}_x) \not> r.$$

Hence, in view of the fact that $\mathbf{M}_x$ has a positive column α_j with entries $a_{kj}(x)$, $k=1, 2, \ldots, n$, which exceed r, it follows that for any $k=1, 2, \ldots, n$

$$a'_{kj}(x) \not< d_c.$$

Consequently, the j-th column of $\mathbf{M}'_x$ contains only positive entries, and moreover

$$\min(a'_{1j}(x), a'_{2j}(x), \ldots, a'_{nj}(x)) \not< d_c.$$

This proves our assertion. In addition, the system of matrices $[D'_s]$ determines at least one number

$$\delta'_c \not< d_c. \tag{67}$$

Now suppose that z is a word over Σ, of length $l(z)=\mu h$. By Theorem 3 (§3) and formulas (56), (67), we have

$$\varkappa(\mathbf{M}_z) \not> \frac{\delta}{6l}; \tag{68}$$

$$\varkappa(\mathbf{M}'_z) \not> \frac{\delta}{6l}. \tag{69}$$

Consequently, by (20) (§4), (68) and (69), we obtain for any word $y=xz$, where $l(z)=\mu h$,

$$\omega(\mathbf{M}_y-\mathbf{M}_z) < \frac{\delta}{6l}; \tag{70}$$

$$\omega(\mathbf{M}'_y-\mathbf{M}'_z) < \frac{\delta}{6l}. \tag{71}$$

It is also easy to see (see (57), (58), (66)) that for any word y of length $l(y) \leqslant \mu h$

$$\omega(\mathbf{M}_y-\mathbf{M}'_y) \not> \frac{\delta}{6l}. \tag{72}$$

It now follows that at once from (70), (71), (72) and formulas (18), (19) in §4 that, for any word $y=xz$, $l(z)=\mu h$,

$$\omega(\mathbf{M}_y-\mathbf{M}'_y) \not> \omega(\mathbf{M}_{xz}-\mathbf{M}_z)+\omega(\mathbf{M}_z-\mathbf{M}'_z)+$$
$$+\omega(\mathbf{M}'_z-\mathbf{M}'_{xz}) < \frac{\delta}{2l}.$$

Hence, via (72), we see that for any word y over Σ,

$$|p_{\mathcal{A}}(y) - p_{\mathcal{A}'}(y)| \ngtr \frac{\delta}{2}. \tag{73}$$

Using (55), (73), we obtain

$$|p_{\mathcal{A}'}(y) - \lambda| \nless \frac{\delta}{2}. \tag{74}$$

Formulas (55), (73) and (74), taken together, show that any FSA $\mathcal{A}'$ satisfying condition (1) with

$$\varepsilon = \frac{\min\left(r, \frac{\delta}{6l}\right)}{(\mu h - 1)\frac{n}{2} + 1},$$

represents the event $T(\mathcal{A}, \lambda)$ with isolated cut-point λ, i. e.,

$$T(\mathcal{A}, \lambda) \approx T(\mathcal{A}', \lambda).$$

This completes the proof of Theorem 6.

§7. POSSIBLE GENERALIZATIONS AND UNSOLVABLE DECISION PROBLEMS

We now consider a natural generalization of quasidefinite finite stochastic automata, and show that the new automata possess many of the typical properties of quasidefinite FSA. To simplify matters, we shall examine only FSA whose input alphabet Σ contains exactly two letters. All the basic definitions and results carry over without difficulty to the case of n input letters.

Let S_n be the class of stochastic $n \times n$ matrices each of whose columns contains exactly one 1. By a theorem of elementary algebra, these matrices form a group under matrix multiplication, which is isomorphic to the group of permutations on n objects and therefore contains exactly $n!$ elements.

Definition 7. Let $\mathbf{A}$ and $\mathbf{B}$ be stochastic $n \times n$ matrices. We shall say that $\{\mathbf{A}, \mathbf{B}\}$ is a *generalized quasidefinite system* if one of the following conditions is fulfilled:

(1) $\{\mathbf{A}, \mathbf{B}\}$ is a quasidefinite system;

(2) there exist h quasidefinite systems of stochastic $n_i \times n_i$ matrices $\{\mathbf{A}_i, \mathbf{B}_i\}$, $i = 1, 2, \ldots, h$, such that for some $\mathbf{T} \in S_n$

$$\mathbf{TAT}^{-1} = \mathbf{A}_1 \dotplus \mathbf{A}_2 \dotplus \ldots \dotplus \mathbf{A}_h;$$

$$\mathbf{TBT}^{-1} = \mathbf{B}_1 \dotplus \mathbf{B}_2 \dotplus \ldots \dotplus \mathbf{B}_h.$$

(The symbol $\dotplus$ denotes the direct sum operation.)

The following properties are trivial corollaries of the appropriate theorems for quasidefinite systems.

Property 1. There is a constructive procedure that decides, for any given system of stochastic $n\times n$ matrices whose entries a satisfy the condition $a=0 \vee a\neq 0$, whether or not the system is generalized quasidefinite.

Property 2. In the general case, the set of generalized quasidefinite systems has an unsolvable decision problem.

Since Definition 7 is of course strongly dependent on the notion of a quasidefinite system, the following theorem may prove useful in tests for generalized quasidefinite systems.

Theorem 7. A stochastic $n\times n$ matrix $\mathbf{M}$, $n\geqslant 2$, has a positive column if

$$\varkappa(\mathbf{M})<\frac{1}{n-1}. \tag{75}$$

Proof. It follows at once from (75) that there is a natural number m with the property

$$\frac{1}{n-1}-\varkappa(\mathbf{M})>\frac{1}{m}. \tag{76}$$

Let $a_{ij}(k)$ denote the k-th term in a sequence of rational numbers defining the entry a_{ij} of $\mathbf{A}$. Suppose that for $k=k'$

$$\forall i \forall j \forall k\Big((k\geqslant k')\supset\Big(|a_{ij}(k')-a_{ij}(k)|<\frac{1}{4m}\Big)\Big), \quad 1\leqslant i,\ j\leqslant n.$$

Since the numbers $a_{ij}(k')$ are rational,

$$\forall i \forall j\Big(\Big(a_{ij}(k')\geqslant\frac{1}{n-1}-\frac{1}{2m}\Big)\vee\Big(a_{ij}(k')<\frac{1}{n-1}-\frac{1}{2m}\Big)\Big).$$
$$1\leqslant i,\ j\leqslant n.$$

Consequently, either there exist i and j, $i=i'$, $j=j'$, such that

$$a_{i'j'}(k')\geqslant\frac{1}{n-1}-\frac{1}{2m}, \tag{77}$$

or for all i, j, $1\leqslant i, j\leqslant n$,

$$a_{ij}(k')<\frac{1}{n-1}-\frac{1}{2m}.$$

By (77), we have

$$\frac{1}{n-1}-\frac{3}{4m}\ngtr a_{i'j'}.$$

Together with (76), this shows that for any f, $1\leqslant f\leqslant n$,

$$\frac{1}{m} < \frac{1}{n-1} - (a_{i'j'} - a_{jj'}) \ngtr a_{jj'} + \frac{3}{4m}.$$

Thus $\frac{1}{4m} \ngtr a_{jj'}$, and so the j'-th column is strictly positive.

Now let

$$\forall i \forall j \left(a_{ij}(k') < \frac{1}{n-1} - \frac{1}{2m} \right), \quad 1 \leqslant i,\ j \leqslant n.$$

We claim that in this case all the entries of **A** are positive. Indeed, by our assumption

$$\forall i \forall j \left(a_{ij} \ngtr \frac{1}{n-1} - \frac{1}{4m} \right), \quad 1 \leqslant i,\ j \leqslant n.$$

Consequently, if $a_{i^*j^*} = 0$ for $i = i^*$, $j = j^*$, then

$$\sum_{j=1}^{n} a_{i^*j^*} \ngtr (n-1) \left(\frac{1}{n-1} - \frac{1}{4m} \right) = 1 - \frac{n-1}{4m} < 1.$$

This of course shows that $a_{i^*j^*} \neq 0$. But since $a_{i^*j^*} \nless 0$, it follows that $a_{i^*j^*} > 0$, and the proof is complete.

Examination of the matrix $\begin{pmatrix} 1 & 0 \\ 0 & 1 \end{pmatrix}$ shows that Theorem 1 cannot be sharpened — the bound $\frac{1}{n-1}$ in (75) is best possible.

Definition 8. An FSA $\mathcal{A} = [\Sigma, S, M_m^n, \pi_0, \alpha]$ is said to be *generalized quasidefinite* if M_m^n is a generalized quasidefinite system.

Property 3. If the system M_m^n of the automaton $\mathcal{A}$ is generalized quasidefinite, there exist a natural number $\mu > 0$ and a rational number δ, $0 < \delta \leqslant 1$, such that for any pair of words x, y over Σ

$$\omega(\pi_0 \mathbf{M}_{yx} - \pi_0 \mathbf{M}_x) \ngtr (1-\delta)^{E\left(\frac{l(x)}{\mu}\right)-1}.$$

The proof will be omitted, since it should present the reader with no difficulties. We shall also omit the proofs of Theorems 8 and 9 below.

*Theorem 8. If λ is an isolated cut-point for a generalized quasidefinite FSA $\mathcal{A}$, then the event $T(\mathcal{A}, \lambda)$ is definite.**

Theorem 9. A generalized quasidefinite FSA is stable relative to any isolated cut-point.

This theorem fails to hold if we replace stability by strong stability, as evidenced by the following example.

Let $\mathcal{A} = [\Sigma, S, M_2^3, \pi_0, \alpha]$ be an FSA with

$$\pi_0 = \left(\frac{1}{3}, \frac{1}{3}, \frac{1}{3} \right);$$

* The converse of this theorem is also valid — any definite event may be represented in some generalized quasidefinite FSA. This follows directly from the representability of definite events by quasidefinite FSA (Perles, Rabin and Shamir, 1963). But since quasidefinite FSA are strongly stable, this implies (in the constructive sense) Rabin's assertion that any definite event is representable by an actual FSA.

$$\alpha = \begin{pmatrix} 1 \\ 0 \\ 1 \end{pmatrix};$$

$$\mathbf{M}_{\sigma_1} = \begin{pmatrix} 1 & 0 & 0 \\ 0 & \frac{4}{5} & \frac{1}{5} \\ 0 & \frac{4}{5} & \frac{1}{5} \end{pmatrix};$$

$$\mathbf{M}_{\sigma_2} = \begin{pmatrix} 1 & 0 & 0 \\ 0 & \frac{1}{2} & \frac{1}{2} \\ 0 & \frac{1}{2} & \frac{1}{2} \end{pmatrix}.$$

The number $\lambda = \frac{1}{2}$ is an isolated cut-point for this FSA. However, it is not an isolated cut-point for the automaton $\mathscr{A}' = [\Sigma, S, M'^3_2, \pi_0, \alpha]$ with

$$\mathbf{M}'_{\sigma_1} = \mathbf{M}_{\sigma_1}; \qquad \mathbf{M}'_{\sigma_2} = \begin{pmatrix} 1-\varepsilon & \varepsilon & 0 \\ 0 & \frac{1}{2} & \frac{1}{2} \\ 0 & \frac{1}{2} & \frac{1}{2} \end{pmatrix}, \qquad \varepsilon > 0.$$

Rabin (1966) considers various decision problems for finite stochastic automata, among these the problem of an effective procedure telling whether or not a cut-point is isolated. Since this problem is generally speaking meaningless for the classical FSA concept, the author proposes confining attention to automata whose transition probabilities are effective real numbers, specifically, rational numbers. Strictly speaking, this formulation of the problem still lacks rigor, and the necessary modifications lead quite naturally to an examination of the problem in the constructive theory of FSA. We now proceed to do this, employing certain ideas of Metra and Smilgais (1968a).*

Let c be any CRN such that $0 < c < 1$. Let $\mathscr{A}^c$ denote the FSA $[\Sigma, S, M_2^2, \pi_0, \alpha]$ with

$$\pi_0 = (1\ \ 0); \qquad \alpha = \begin{pmatrix} 0 \\ 1 \end{pmatrix};$$

$$\mathbf{M}_{\sigma_1} = \begin{pmatrix} 1 & 0 \\ \frac{1}{2} & \frac{1}{2} \end{pmatrix}; \qquad \mathbf{M}_{\sigma_2} = \begin{pmatrix} 1-c & c \\ 0 & 1 \end{pmatrix}.$$

Lemma 3. *The FSA $\mathscr{A}^c$ has an isolated cut-point λ if and only if $c > \frac{1}{2}$.*

* This paper was written in the spirit of the classical theory, so that its results do not carry over directly to the constructive theory.

Proof. Let $Q_c(\lambda)$ denote the statement that λ is an isolated cut-point for the FSA $\mathscr{A}^c$. We shall prove that

$$\left(c>\frac{1}{2}\right)\supset \exists\,\lambda Q_c(\lambda). \tag{78}$$

By the definition of $\mathscr{A}^c$,

$$\pi_0 \mathbf{M}_x \mathbf{M}_{\sigma_1} \alpha = 0a_{11}(x) + \frac{1}{2}a_{12}(x) \not> \frac{1}{2};$$

$$\pi_0 \mathbf{M}_x \mathbf{M}_{\sigma_2} \alpha = ca_{11}(x) + 1a_{12}(x) \not< c(a_{11}(x) + a_{12}(x)) = c.$$

Consequently, for all words x over Σ and $\lambda = \frac{1}{2} + \frac{c-\frac{1}{2}}{2}$, we have

$$\left(p_{\mathscr{A}^c}(x) \not> \lambda - \frac{c-\frac{1}{2}}{2}\right) \vee \left(p_{\mathscr{A}^c}(x) \not< \lambda + \frac{c-\frac{1}{2}}{2}\right).$$

This proves (78).

It is a little more difficult to prove

$$\exists\,\lambda Q_c(\lambda) \supset \left(c>\frac{1}{2}\right). \tag{79}$$

We shall prove the following statement. If $c \not> \frac{1}{2}$, then for any natural number k the input words of the automaton $\mathscr{A}^c$ of length k may be indexed in such a way that if x_i is the i-th word the entries of the matrix $\mathbf{M}_{x_i}$ satisfy the conditions

$$\begin{cases} a_{12}(x_1) = 0; \\ a_{22}(x_{2^k}) = 1; \\ a_{12}(x_i) < a_{22}(x_i); \\ a_{12}(x_{i+1}) \not> a_{22}(x_i). \end{cases} \tag{80}$$

(Although the third formula is independent of how the words are indexed, its inclusion here involves no logical contradictions.)

If $k=1$, conditions (80) follow directly from the definition of $\mathscr{A}^c$, with $x_1=\sigma_1$, $x_2=\sigma_2$. Suppose now that the statement is true for $k=k^*$. We shall show that it is true for $k=k^*+1$ too. To simplify the notation, we denote $\mathbf{M}_{x_i}$ by $\mathbf{B}$ and $\mathbf{M}_{\sigma_1}\mathbf{B}$, $\mathbf{M}_{\sigma_2}\mathbf{B}$ by $\mathbf{C}$, $\mathbf{D}$, respectively. Then $b_{12}=c_{12}$; $b_{22}=d_{22}$; $c_{12}<c_{22}$; $d_{12}<d_{22}$; $c_{22}\not< d_{12}$. We shall prove only the last formula:

$$d_{12} = (1-c)b_{12} + cb_{22} = b_{12} + c(b_{22} - b_{12});$$

$$c_{22} = \frac{1}{2}b_{12} + \frac{1}{2}b_{22} = b_{12} + \frac{1}{2}(b_{22} - b_{12}).$$

It follows from the inductive hypothesis that $b_{22} - b_{12} > 0$. Thus we must have $d_{12} \not> c_{22}$.

It is now readily seen that the statement is true for $k=k^*+1$ and so for every $k\geqslant 1$. Hence, since

$$\varkappa(\mathbf{M}_{x_i}) \not> (1-c)^k$$

(this follows directly from the definition of $\mathcal{A}^c$ and the properties of quasi-definite systems of stochastic matrices), we obtain

$$\forall\lambda\,\forall l\left[(0<\lambda<1\&0<l)\supset\exists x\left(|p_{\mathcal{A}^c}(x)-\lambda|<\frac{1}{l}\right)\right].$$

We have thus proved

$$\left(c\not>\frac{1}{2}\right)\supset\neg\exists\lambda Q_c(\lambda).$$

Applying appropriate logical rules to (81), we see that

$$\exists\lambda Q_c(\lambda)\supset\neg\neg\left(c>\frac{1}{2}\right). \tag{82}$$

Hence

$$\exists\lambda Q_c(\lambda)\supset\left(c>\frac{1}{2}\right).$$

This completes the proof of Lemma 3.

Theorem 10. There is no algorithm that decides for any given FSA whether or not it has an isolated cut-point.

Proof. Let c^* denote the numbers defined by

$$c^*=\frac{1}{2}+\frac{d}{2},\quad d\in\{B_g\}, \tag{83}$$

and consider the class of FSA containing all the automata $\mathcal{A}^{c^*}$. By Lemma 3,

$$\forall c^*\left[\left(\exists\lambda Q_{c^*}(\lambda)\vee\neg\exists\lambda Q_{c^*}(\lambda)\right)\leftrightarrow\left(c^*>\frac{1}{2}\vee c^*\not>\frac{1}{2}\right)\right].$$

Hence, by (83), we obtain

$$\forall c^*\left[(\exists\lambda Q_{c^*}(\lambda)\vee\neg\exists\lambda Q_{c^*}(\lambda))\leftrightarrow(d>0\vee d\not>0)\right]. \tag{84}$$

Using the properties of the set of numbers d, we readily deduce the required formula from (84):

$$\neg\forall c^*(\exists\lambda Q_{c^*}(\lambda)\vee\neg\exists\lambda Q_{c^*}(\lambda)). \tag{85}$$

Note that Theorem 10 does not solve the problem in the realm of rational numbers, i. e., it remains open when the transition probabilities of the FSA are rational numbers. One indication of this is the fact that the numbers d are rational only in the classical sense; they cannot be proved rational by constructive means.

NOTES

The classical analogs of Theorems 2, 4 and 6 were proved by Paz and published earlier than the corresponding constructive results, which we established independently of Paz (Lorenz, 1967, 1968b). The most important results of Paz (1966a) were then rediscovered by Kochkarev.

Our definition of FSA is essentially the same as the classical definitions of Rabin and Paz. Carlyle, Bukharaev and Starke used another, more general definition, which includes in a natural way the concepts of stochastic automata with enumerable sets of input and internal states. Moreover, in their definitions the transition probability also depends on the output.

Representation of events by stochastic automata has been considered in the traditional framework by Metra, Starke, Bukharaev, Kochkarev, Salomaa, Turakainen, and others. The most profound investigation is in our opinion that of Turakainen (1968), who considers the generalized automata first defined by Page (1966). Starke (1965, 1966a, b) also studied another definition of representability of events, suggested by Thiele — representation of stochastic events. The gist of this approach is as follows: any set of words may be defined through a suitable characteristic function $f(x)$; instead of characteristic functions of the traditional type $((f(x)=0)\vee(f(x)=1))$, Starke considers functions $f(x)$ whose values satisfy the condition $0\leqslant f(x)\leqslant 1$. In a certain sense, any such function may be viewed as a stochastic event, represented in an automaton $\mathcal{A}$ by a condition of the type

$$\forall x(f(x)=p_{\mathcal{A}}(x)).$$

In the classical literature, much attention is devoted to various questions concerning equivalence of automata. Most representative in this respect is the highly interesting paper of Carlyle (1963). Similar problems have been studied by Bacon (1964), Starke (1968, 1969b), Bukharaev (1964a, b, 1966), Paz (1966b), Nawrotzki (1966), and others. However, it should be noted that none of these results carries over to the constructive case. Nevertheless, it is quite legitimate to formulate constructive proofs when the basic numerical parameters are rational numbers. This reservation is certainly valid for Carlyle's fundamental theorem on reduced forms of FSA. Not without interest for the constructivist is the classically conceived paper of Zühlke (1969), who formulates the problem of ε-substitutable FSA; it is precisely in this area that the constructive approach has something to offer outside the rational domain.

There is comparatively little literature on stability of FSA (see §5, Definition 6); this is probably due to the difficulty of the problems involved. It would be interesting to derive conditions for an FSA to be unstable, based on the ideas behind Kesten's example (see Arbib, 1969). We have introduced certain corrections to the original definition of stability, so that Definition 6 differs from our earlier one (Lorenz, 1967, 1968).

Stability problems are intimately connected with the problem of determining FSA with isolated cut-points. Both constructivist and classicist are interested in tests for the existence of an isolated cut-point. Some work in this direction has been carried out by Metra and Smilgais (1968a) and Podnieks (1970). Metra and Smilgais present an effective test for the existence of isolated cut-points of FSA with two internal states and rational transition probabilities.

Chapter VII

ABSTRACT SYNTHESIS OF FINITE STOCHASTIC AUTOMATA

Unlike the last chapter, which was concerned with analysis of FSA, this chapter will be devoted to the design of FSA with prescribed properties. Since these constructions concern only the components of the abstract definition of the automaton (specifically, the input and internal state of alphabets, system of transition matrices, initial distribution of internal states and vector of final states), they are usually subsumed under the heading "abstract synthesis." This topic should not be confused with structural synthesis, where one is actually interested in the construction of certain labeled graphs.

From the classical standpoint, problems of abstract synthesis are in essence problems of analysis, since there is no demand for a constructive method producing the automaton. In the absence of such constructive procedures, however, abstract synthesis does not yield valid constructions of concrete automata.

§1. FUNDAMENTAL REDUCTION THEOREM

It is readily observed that Rabin's theorem (Rabin, 1963) to the effect that every FSA $\mathfrak{A}$ with an isolated cut-point λ represents a regular event is just another instance of a pure existence theorem, since its proof provides no method for construction of an FDA representing $T(\mathfrak{A}, \lambda)$. As a matter of fact, the classical conception of an FSA does not even permit determination of an effective construction procedure. This is easily seen from the following example.

Consider the (classical) FSA $\mathcal{A}=[\Sigma, S, \mathbf{M}_2^2, \pi_0, \alpha]$ with $\pi_0=(1, 0)$,

$$\mathbf{M}_{\sigma_1}=\begin{pmatrix} 1-\operatorname{sign} d & \operatorname{sign} d \\ 1-\operatorname{sign} d & \operatorname{sign} d \end{pmatrix}, \quad d\in\{B_g\};$$

$$\mathbf{M}_{\sigma_2}=\begin{pmatrix} 0 & 1 \\ 1 & 0 \end{pmatrix};$$

$$\alpha=\begin{pmatrix} 0 \\ 1 \end{pmatrix}.$$

Each FSA thus defined has an isolated cut-point $\lambda=\frac{1}{2}$, and in addition it is an FDA in the classical sense. But there can be no effective procedure deciding for any d which of the components of the disjunction

$$\left(\sigma_1 \in T\left(\mathcal{A}, \frac{1}{2}\right)\right) \vee \left(\sigma_1 \notin T\left(\mathcal{A}, \frac{1}{2}\right)\right)$$

is true.

In spite of this counterexample, Rabin's reduction theorem is not without value for our purposes, since the ideas embodied in its proof may be used to prove the analogous theorem for the constructive theory of FSA.

Theorem 1. If λ is an isolated cut-point for an FSA $\mathcal{A}$, then $T(\mathcal{A}, \lambda)$ is a regular event.

Proof. Let $\mathcal{A} = [\Sigma, S, \mathbf{M}_m{}^n, \pi_0, \alpha]$ be an FSA with cut-point λ, such that for some rational number $\delta > 0$

$$\forall x(|p_{\mathcal{A}}(x) - \lambda| \not< \delta). \qquad (1)$$

We fix a natural number k, and divide the set of all input words x (including the empty word) of length $l(x) \leqslant k$ into classes $B_k{}^e$, where e is a word of length $k' \leqslant k$ over Σ: $x \in B_k{}^e$ if and only if, for any z over Σ with $l(z) \leqslant k$,

$$(p_{\mathcal{A}}(xz) > \lambda + \delta) \leftrightarrow (p_{\mathcal{A}}(ez) > \lambda + \delta). \qquad (2)$$

Denoting the number of distinct classes $B_k{}^e$ by φ_k, we shall show that

$$\varphi_k \leqslant \left(1 + \frac{n}{2\delta}\right)^n. \qquad (3)$$

Indeed, suppose that x and x' belong to different classes; then we may effectively determine a word z such that

$$p_{\mathcal{A}}(xz) > \lambda + \delta, \quad \text{but} \quad p_{\mathcal{A}}(x'z) < \lambda - \delta$$

or, conversely,

$$p_{\mathcal{A}}(xz) < \lambda - \delta, \quad \text{but} \quad p_{\mathcal{A}}(x'z) > \lambda + \delta.$$

Denoting the entries of the matrix $\pi_0 \mathbf{M}_x$ by c_i, $i = 1, 2, \ldots, n$, and those of $\pi_0 \mathbf{M}_{x'}$ by c'_i, $i = 1, 2, \ldots, n$, we can prove that

$$\sum_{i=1}^{n} |c_i - c'_i| > 2\delta. \qquad (4)$$

In fact, the matrix

$$\mathbf{M}_z \alpha = \begin{pmatrix} d_1 \\ d_2 \\ \cdots \\ \cdots \\ d_n \end{pmatrix}$$

has only nonnegative entries, none exceeding unity, and since

$$p_{\mathcal{A}}(xz) = \pi_0 \mathbf{M}_x \mathbf{M}_z \alpha = \sum_{i=1}^{n} c_i d_i,$$

$$p_{\mathcal{A}}(x'z) = \pi_0 \mathbf{M}_{x'} \mathbf{M}_z \alpha = \sum_{i=1}^{n} c'_i d_i,$$

it follows that

$$\sum_{i=1}^{n} |c_i - c'_i| \not< \sum_{i=1}^{n} d_i |c_i - c'_i| \not< |p_{\mathcal{A}}(xz) - p_{\mathcal{A}}(x'z)| > 2\delta.$$

From (4) it follows that

$$\max_{i=1,2,\ldots,n} |c_i - c'_i| > \frac{2\delta}{n}. \tag{5}$$

We now define a new metric in euclidean n-space, with respect to which the distance between points $t = (b_1, b_2, \ldots, b_n)$ and $t' = (b'_1, b'_2, \ldots, b'_n)$ is

$$\mathrm{P}(t, t') = \max_{i=1,2,\ldots,n} |b_i - b'_i|. \qquad (6)^*$$

It is easy to see that the ε-neighborhood of a point t_0 in the space, i. e., the set of points defined by the condition $\mathrm{P}(t_0, t) < \varepsilon$, is represented geometrically by an n-dimensional cube with edge 2ε and volume $(2\varepsilon)^n$. It follows from (5) and (6) that the $\frac{\delta}{n}$-neighborhoods of the points $\tilde{t} = (c_1, c_2, \ldots, c_n)$ and $\tilde{t}' = (c'_1, c'_2, \ldots, c'_n)$ are disjoint.

Consequently, if the number of distinct classes B_k^e is φ_k, we can find an n-dimensional cube with edge $1 + \frac{2\delta}{n}$ in which there are φ_k n-dimensional cubes with edge $\frac{2\delta}{n}$. It follows that

$$\varphi_k \left(\frac{2\delta}{n}\right)^n \leqslant \left(1 + \frac{2\delta}{n}\right)^n,$$

and this implies (3).

We now let $n' = E\left(\left(1 + \frac{n}{2\delta}\right)^n\right)$ and set $n' - \varphi_{n'} = h$. It follows from (3) that there exists a natural number k, $n' \leqslant k \leqslant n' + h$, such that each class B_k^e contains at least one word x of length $l(x) < k$.

Let k' be the smallest such number; we define an FDA $\mathcal{A}' = [\Sigma, S', \mathbf{M}'^{\varphi_{k'}}_m, \pi'_0, \alpha]$ as follows. Take the $\varphi_{k'}$ pairwise distinct classes $B^e_{k'}$, index them from 1 to $\varphi_{k'}$ inclusive, and denote them for brevity by s'_i. We shall assume that s'_1 is the class $B^e_{k'}$ containing the empty word. Now set $S' = \{s'_1, s'_2, \ldots, s'_{\varphi_{k'}}\}$. The final state vector α' is defined thus: $\alpha'_i = 1$ if and only if the class $B^e_{k'}$ corresponding to s'_i contains a word $x \in T(\mathcal{A}, \lambda)$.

* This metric was considered, in the two-dimensional case, by Heyting (1956).

We now define the matrices $\mathbf{M}'_{\sigma_j} = (a_{il}(\sigma_j))$: $a_{il}(\sigma_j) = 1$ if there exists a word x, $l(x) < k'$, in the class $B^e_{k'}$ corresponding to s'_i, such that the word $x\sigma_j$ belongs to the class $B^e_{k'}$ corresponding to s'_l; $a_{il}(\sigma_j) = 0$ otherwise.

We claim that the FDA $\mathcal{A}'$ represents the event $T(\mathcal{A}, \lambda)$, if $p'_1 = 1$. Our goal will be achieved if we can show that for all words x

$$(p_{\mathcal{A}'}(x) = 1) \leftrightarrow (p_{\mathcal{A}}(x) > \lambda). \tag{7}$$

Suppose on the contrary that there is a word x^* satisfying the condition

$$((p_{\mathcal{A}'}(x^*) = 1) \& (p_{\mathcal{A}}(x^*) < \lambda)) \vee ((p_{\mathcal{A}}(x^*) > \lambda) \& (p_{\mathcal{A}'}(x^*) = 0). \tag{8}$$

By the construction of $\mathcal{A}'$,

$$l(x^*) > 2k'; \qquad \varphi_{l(x^*)} > \varphi_{k'}.$$

But since $\varphi_{l(x^*)}$ satisfies (3), there exists a natural number k^*, $l(x^*) \leqslant k^* \leqslant \leqslant l(x^*) + (n' - \varphi_{l(x^*)})$, such that any class $B^e{}_{k^*}$ contains at least one word x of length $l(x) < k^*$. Consequently, we can construct an FDA $\mathcal{A}^* = [\Sigma, S^*, \mathbf{M}^*{}_m{}^{\varphi_{k^*}}, \pi^*{}_0, \alpha^*]$ with φ_{k^*} internal states, such that

$$\forall x((l(x) \leqslant 2k^*) \supset ((p_{\mathcal{A}^*}(x) = 1) \leftrightarrow (p_{\mathcal{A}}(x) > \lambda))).$$

In view of our assumption on x^*, this implies

$$\neg((p_{\mathcal{A}'}(x^*) = 1) \leftrightarrow (p_{\mathcal{A}^*}(x^*) = 1)). \tag{9}$$

But since $\max(\varphi_{k'}, \varphi_{k^*}) \leqslant n'$, it follows that $\mathcal{A}'$ and $\mathcal{A}^*$ may be identified by multiple experiments of length $2n'$, i. e., the description is completely defined by the values of $p_{\mathcal{A}'}(x)$ and $p_{\mathcal{A}^*}(x)$ for words x of length $l(x) \leqslant 2n'$. Now for words of this length the behavior of the FDA $\mathcal{A}'$ is exactly the same as that of $\mathcal{A}^*$. Consequently, $\mathcal{A}'$ and $\mathcal{A}^*$ are equivalent, and this contradicts (9). Thus our assumption concerning the existence of x^* is false and we may write

$$\neg\exists x(((p_{\mathcal{A}'}(x) = 1) \& (p_{\mathcal{A}}(x) < \lambda)) \vee ((p_{\mathcal{A}}(x) > \lambda) \& (p_{\mathcal{A}'}(x) = 0))).$$

Hence, by the laws of constructive logic and formula (1), we finally get

$$\forall x(((p_{\mathcal{A}'}(x) = 1) \supset (p_{\mathcal{A}}(x) > \lambda)) \& ((p_{\mathcal{A}}(x) > \lambda) \supset (p_{\mathcal{A}'}(x) = 1))).$$

This proves (8) and thereby completes the proof of Theorem 1.

§2. NORMAL REPRESENTATION THEOREM

In practical use of finite automata as data-processing devices, one usually demands more than just representation of a certain event with cut-point λ. One is most frequently interested in representations which might be called normal, i.e., representations with an isolated cut-point $\lambda=\frac{1}{2}$. This is the predominant standpoint of von Neumann (1956) and Moore (1956), as evidenced by the sections of their papers devoted to design of combinational and relay switching networks with reliability $p>\frac{1}{2}$, to replace unreliable components and relays. The same position is held by Kovalenko, who points out the questionable value of representation of nonregular events by stochastic automata with nonisolated cut-points. One is thus naturally faced with the problem of normal representation of an event U, provided of course that U is at all representable in some FSA with isolated cut-point. The answer to the question, which is indeed positive, follows directly from the reduction theorem, but this necessitates an excursion into the realm of FDA, which is not always advisable in view of practical limitations. There is therefore some point in seeking a less "deterministic" answer to the question than the reduction theorem.

Since the traditional terminology is not really expressive of the state of affairs, we introduce a few new definitions.

Definition 1. Let $[G_1, G_2]$ be an ordered pair of sets of words over an alphabet. Σ. We shall call $[G_1, G_2]$ an information medium (IM) over Σ if G_1 and G_2 are disjoint sets with solvable decision problems.

Definition 2. Let $\mathscr{A}=[\Sigma, S, M_m{}^n, \pi_0, \alpha]$ be an FSA, $[G_1, G_2]$ an information medium over Σ, $\delta>0$ a rational number and λ, $0<\lambda<1$, a CRN. We shall say that $\mathscr{A}$ represents the IM $[G_1, G_2]$ with cut-point λ and radius of isolation δ if

$$\underset{x\in G_1}{\forall} x(p_{\mathscr{A}}(x)\not<\lambda+\delta)\;\&\;\underset{x\in G_2}{\forall} x(p_{\mathscr{A}}(x)\not>\lambda-\delta).$$

An information medium $[G_1, G_2]$ is said to be finite if G_1 and G_2 are finite sets. If $\mathscr{A}$ represents an IM with cut-point $\lambda=\frac{1}{2}$ and radius of isolation δ, we shall also say that $\mathscr{A}$ represents the IM with parameter δ.

Theorem 2. If an FSA $\mathscr{A}=[\Sigma, S, \mathbf{M}_m{}^n, \pi_0, \alpha]$ represents an IM $[G_1, G_2]$ with cut-point λ and radius of isolation δ, then for any rational number δ', $0<\delta'<\frac{1}{2}$, one can construct an FSA $\mathscr{A}'$ representing $[G_1, G_2]$ with parameter δ', with the property that

$$\underset{x\in(G_1\cup G_2)}{\forall} x(((p_{\mathscr{A}}(x)>0)\leftrightarrow(p_{\mathscr{A}'}(x)>0))\,\&\,((p_{\mathscr{A}}(x)<1)\leftrightarrow \leftrightarrow p_{\mathscr{A}'}(x)<1))). \tag{10}$$

Proof. For any nonnegative rational number r, we let $]r[$ denote the natural number such that $0\leqslant]r[-r<1$.

To simplify matters, we assume that λ is a rational number,* and find the smallest natural number k such that

$$]k\lambda[\neq k\lambda;$$

$$\frac{\max\{(\lambda-\delta)(1-\lambda+\delta),\ (\lambda+\delta)(1-\lambda-\delta)\}}{k\delta^2}<\varepsilon=\frac{1}{2}-\delta'. \tag{11}$$

* Otherwise, one can construct rational numbers $\tilde{\lambda}$ and $\tilde{\delta}$ such that $\mathscr{A}$ represents the IM with cut-point $\tilde{\lambda}$ and radius of isolation $\tilde{\delta}$.

It is clear that the required number is either

$$k=\left]\frac{\max\{(\lambda-\delta)(1-\lambda+\delta),\ (\lambda+\delta)(1-\lambda-\delta)\}}{\varepsilon\delta^2}\right[$$

or

$$k=\left]\frac{\max\{(\lambda-\delta)(1-\lambda+\delta),\ (\lambda+\delta)(1-\lambda-\delta)\}}{\varepsilon\delta^2}\right[+1.$$

Denote the required value of k by k'. Suppose that all words of length k' over the alphabet S are ordered lexicographically and indexed from 1 to $n^{k'}$. We let s'_i denote the i-th word in this ordering and define an alphabet S' by $S'=\{s'_1, s'_2, \ldots, s'_\nu\}$, where $\nu=n^{k'}$.

Now let $l^*(f)$ denote the number of occurrences of final states s_i in a word f over S. For example, if $n=3$, the only final states of $\mathcal{A}$ are s_1 and s_2, and $f=s_1s_1s_3s_2s_3s_2$, then $l^*(f)=4$. The final state vector is defined by

The final state vector is defined by

$$((\alpha'_i=1)\leftrightarrow(l^*(s'_i)\geqslant]k'\lambda[)). \tag{12}$$

To define π'_0, we let p'_i and p_j denote respectively the i-th ($i=1, 2, \ldots, \nu$) and j-th ($j=1, 2, \ldots, n$) components of π'_0 and π_0. If $s'_i=s_{i_1}s_{i_2}\ldots s_{i_{k'}}$, we set

$$p'_i=p_{i_1}\,p_{i_2}\ldots p_{i_{k'}}. \tag{13}$$

Now let

$$s'_t=s_{t_1}\,s_{t_2}\ldots s_{t_{k'}};\qquad s'_u=s_{u_1}\,s_{u_2}\ldots s_{u_{k'}}.$$

We define the entry $a'_{tu}(\sigma_i)$ of the matrix $\mathbf{M}'_{\sigma_i}$ by

$$a'_{tu}(\sigma_i)=a_{t_1u_1}(\sigma_i)\,a_{t_2u_2}(\sigma_i)\ldots a_{t_{k'}u_{k'}}(\sigma_i). \tag{14}$$

We omit the details of the simple verification that the system $[\Sigma,\ S',\ \mathbf{M}_m{}'^{\nu},\ \pi'_0,\ \alpha']$ indeed defines an FSA $\mathcal{A}'$.

We claim that every entry $a'_{tu}(x)$ of the matrix $\mathbf{M}'_x$ satisfies

$$a'_{tu}(x)=a_{t_1u_1}(x)\ldots a_{t_{k'}u_{k'}}(x). \tag{15}$$

Indeed, if $l(x)=1$ formula (15) is valid by definition. Supposing it true for every word x of length $l(x)\leqslant\vartheta$, we show that it is true for any x' of length $l(x')=\vartheta+1$. There exist words y and z such that

$$\max(l(y),\ l(z))\leqslant\vartheta;\qquad x'=yz. \tag{16}$$

Since $\mathbf{M}'_{x'}=\mathbf{M}'_y\mathbf{M}'_z$, it follows from (16) that

$$a'_{tu}(x')=\sum_{v=1}^{\nu}a'_{tv}(y)\,a'_{vu}(z)=$$

$$=\sum_{(i_1i_2\ldots i_{k'})=(11\ldots1)}^{(nn\ldots n)}a_{t_1i_1}(y)\cdot\ldots\cdot a_{t_{k'}i_{k'}}(y)\cdot a_{i_1u_1}(z)\cdot\ldots\cdot a_{i_{k'}u_{k'}}(z)=$$

$$=\prod_{k=1}^{k'}\left(\sum_{j=1}^{n} a_{t_kj}(y)a_{ju_k}(z)\right)=\prod_{k=1}^{k'} a_{t_ku_k}(x').$$

Thus $a'_{tu}(x')=a_{t_1u_1}(x')\dots a_{t_{k'}u_{k'}}(x')$. This establishes (15) for any word x.

Let $p_j(x)$ and $p'_i(x)$ denote the corresponding components of the stochastic vectors $\pi_0\mathbf{M}_x$ and $\pi'_0\mathbf{M}'_x$, respectively. We shall show that

$$p'_i(x)=p_{i_1}(x)\dots p_{i_{k'}}(x) \tag{17}$$

for any word x and any i, $1\leqslant i\leqslant \nu$, where $s'_i=s_{i_1}s_{i_2}\dots s_{i_{k'}}$.

By (13) and (15),

$$p'_i(x)=\sum_{t=1}^{\nu}p'_ta'_{ti}(x)=\sum_{(i_1i_2\dots i_{k'})=(11\dots1)}^{(nn\dots n)} p_{j_1}\dots p_{j_{k'}}a_{j_1i_1}(x)\dots a_{j_{k'}i_{k'}}(x)=$$

$$=\prod_{k=1}^{k'}\left(\sum_{j=1}^{n} p_ja_{ji_k}(x)\right)=\prod_{k=1}^{k'} p_{i_k}(x).$$

This proves (17).

Now fix a word x and define a probability field $\Omega'=(S',\mathfrak{P}')$ by

$$\mathfrak{P}'(s'_i)=p'_i(x). \tag{18}$$

By (17), Ω' may be treated as the field of k'-independent trials associated with the probability field $\Omega=(S,\mathfrak{P})$, we have

$$\mathfrak{P}(s_j)=p_j(x). \tag{19}$$

Let β denote a k'-vector with coordinates β_j from the set $\{0, 1\}$. Given β, we define an event B_β over S' by the condition: $s'_i=s_{i_1}s_{i_2}\dots s_{i_{k'}}$ belongs to B_β if and only if

$$\forall i((\alpha_{ij}=1)\leftrightarrow(\beta_j=1)).$$

It is readily verified that

$$p(B_\beta)=(p_{\mathcal{A}}(x))^{\nu_\beta}(1-p_{\mathcal{A}}(x))^{k'-\nu_\beta}, \tag{20}$$

where ν_β is the number of nonzero components of β.

Indeed, if $k'=1$, formula (20) is trivially true. Supposing it true for all $k'\leqslant h$, let us prove it for $k'=h+1$. To this end, we consider two vectors

$$\beta=(\beta_1,\beta_2,\dots,\beta_h);\quad \beta'=(\beta_1,\beta_2,\dots,\beta_{h+1}).$$

To fix ideas, we assume that $\beta_{h+1}=1$. * Then $s'_i=s_{i_1}\dots s_{i_h}s_{i_{h+1}}$ is in $B_{\beta'}$ if and only if

$$(\alpha_{i_{h+1}}=1)\&(s_{i_1}\dots s_{i_h}\in B_\beta).$$

* In the other case the reasoning is analogous.

Hence by (19) and the inductive hypothesis, we obtain

$$p(B_{\beta'}) = \sum_{i=1}^{n} p(B_\beta)\mathfrak{B}(s_i)\alpha_i = p(B_\beta)p_{\mathscr{A}}(x) =$$

$$= (p_{\mathscr{A}}(x))^{\nu_\beta+1}\ (1-p_{\mathscr{A}}(x))^{h-\nu_\beta} = (p_{\mathscr{A}}\ (x))^{\nu_{\beta'}}(1-p_{\mathscr{A}}(x))^{h+1-\nu_{\beta'}}.$$

Thus,

$$p(B_{\beta'}) = (p_{\mathscr{A}}(x))^{\nu_{\beta'}}\ (1-p_{\mathscr{A}}\ (x))^{h+1-\nu_{\beta'}},$$

which proves (20) for arbitrary k'.

Define a random variable $\mathfrak{P}$ over Ω' by $\mathfrak{P}(s'_i) = \nu_\beta$ if $s'_i \in B_\beta$.

Since the events B_β are pairwise mutually exclusive, $\mathfrak{P}$ is well defined. By (18) and (20) the mean $M(\mathfrak{P})$ and variance $D(\mathfrak{P})$ are given by

$$M(\mathfrak{P}) = k'p_{\mathscr{A}}(x); \tag{21}$$

$$D(\mathfrak{P}) = k'p_{\mathscr{A}}(x)\,(1-p_{\mathscr{A}}(x)). \tag{22}$$

We now show that $\mathscr{A}'$

$$\underset{x\in(G_1\cup G_2)}{\forall} x\left(\left|p_{\mathscr{A}'}(x) - \frac{1}{2}\right| > \delta'\right). \tag{23}$$

The proof of this assertion will be divided into two parts:

$$\underset{x\in G_1}{\forall} x\left(p_{\mathscr{A}'}(x) > \frac{1}{2} + \delta'\right); \tag{24}$$

$$\underset{x\in G_2}{\forall} x\left(p_{\mathscr{A}'}(x) < \frac{1}{2} - \delta'\right). \tag{25}$$

If $x \in G_1$, we let d denote the difference $p_{\mathscr{A}}(x) - \lambda - \delta$. By Chebyshev's inequality and formulas (11) and (22), we have

$$p_{\{|\mathfrak{P}-M(\mathfrak{P})|>k'(\delta+d)\}} \dot{\ngtr} \frac{D(\mathfrak{P})}{k'^2(\delta+d)^2} = \frac{(\lambda+\delta+d)\,(1-\lambda-\delta-d)}{k'\,(\delta+d)^2} =$$

$$= \frac{\lambda(1-\lambda) + (1-2\lambda)\,(\delta+d) - (\delta+d)^2}{k'\,(\delta+d)^2} \ngtr$$

$$\ngtr \frac{\max\{\lambda(1-\lambda) + (1-2\lambda)\delta - \delta^2,\ \lambda(1-\lambda) - (1-2\lambda)\delta - \delta^2\}}{k'\delta^2} =$$

$$= \frac{\max\{(\lambda+\delta)\,(1-\lambda-\delta),\ (\lambda-\delta)\,(1-\lambda+\delta)\}}{k'\delta^2} < \varepsilon.$$

Consequently,

$$p_{\{|\mathfrak{P}-M(\mathfrak{P})|>k'(\delta+d)\}} \dot{<} \varepsilon. \tag{26}$$

Obviously (see (21)), the value $\mathfrak{P}=\mathfrak{P}(s'_i)$ satisfies the condition

$$|\mathfrak{P}(s'_i)-M(\mathfrak{P})|<k'(\delta+d),$$

if and only if

$$(\mathfrak{P}(s'_i)>k'(\lambda+2(\delta+d)))\vee(\mathfrak{P}(s'_i)<k'\lambda). \tag{27}$$

Since the values of $\mathfrak{P}$ and the numbers $k'\lambda$ are rational numbers, the probability $p(\{\mathfrak{P}<]k'\lambda[\})$ is well-defined, and by formulas (26) and (27) we have $p(\{\mathfrak{P}<]k'\lambda[\})<\varepsilon$. On the other hand (see (11), (12)), $p(\{\mathfrak{P}<]k'\lambda[\})=$ $=1-p_{\mathcal{A}'}(x)$, whence we obtain $p_{\mathcal{A}'}(x)>1-\varepsilon=\frac{1}{2}+\delta'$. This proves (24).

Now let $x\in G_2$. Then there is a number $d'\nless 0$ such that $p_{\mathcal{A}}(x)=\lambda-(\delta+d')$. By Chebyshev's inequality and formulas (11), (22),

$$p_{\{|\mathfrak{P}-M(\mathfrak{P})|>k'(\delta+d')\}}<\varepsilon. \tag{28}$$

It is easy to verify the formula (see (21))

$$\forall s'_i((|\mathfrak{P}(s'_i)-M(\mathfrak{P})|>k'(\delta+d'))\leftrightarrow(\mathfrak{P}(s'_i)>k'\lambda\vee \\ \vee\mathfrak{P}(s'_i)<k'(\lambda-2(\delta+d')))). \tag{29}$$

It follows from (28) and (29) that

$$p(\{\mathfrak{P}(s'_i)\geqslant]k'\lambda[\})<\varepsilon. \tag{30}$$

But since

$$p(\{\mathfrak{P}(s'_i)\geqslant]k'\lambda[\})=p_{\mathcal{A}'}(x),$$

we see from (30) that

$$p_{\mathcal{A}'}(x)<\frac{1}{2}-\delta'.$$

This proves (25).

Formula (10) follows easily from (12) and (20). The proof is thus complete.

It is readily seen (see (15)) that if $\mathcal{A}$ is a quasidefinite (actual) automaton then the new automaton $\mathcal{A}'$ is also quasidefinite (actual). It is also easy to show that $\mathcal{A}'$ is stable relative to the cut-point $\frac{1}{2}$ if $\mathcal{A}$ is stable relative to λ.

§3. SAVING OF STATES

Frequently, the engineer wishing to guarantee highly reliable transmission of messages along an information channel resorts to codes which permit detection and correction of a certain number of errors (distortions). If the information thus recorded is then processed by some automaton (such as a decoder), the input to the latter will consist only of input signals of the stipulated type. Of course, the use of self-correcting codes does not mean that no "illegal" signal sequences may appear; however — and this is important — the frequency of noninterpretable sequences is always small. Situations thus arise in which certain responses of the automaton may be assured only relative to a certain type of input sequence. In all other cases, i. e., upon the appearance of "illegal" input sequences, the automaton's response may be arbitrary

For a rigorous description of these ideas concerning the operating conditions of automata, one can use one of the most general definitions of FSA, due to Lorenz (1968).

According to this definition, an FSA is a system $[\Sigma, S, \mathbf{M}_m{}^n, \pi_0, \alpha^1, \alpha^2, \ldots \alpha^h]$, where α^i, $1 \leqslant i \leqslant h$, is a column vector

$$\begin{pmatrix} \alpha_1{}^i \\ \alpha_2{}^i \\ \cdots \\ \alpha_n{}^i \end{pmatrix},$$

whose components are either 0 or 1. It is assumed that

$$\alpha^1 + \alpha^2 + \ldots + \alpha^h = \begin{pmatrix} 1 \\ 1 \\ \cdots \\ 1 \end{pmatrix}.$$

Using this FSA concept, we can generalize Definition 2 in a natural manner for $\lambda = \frac{1}{2}$, after suitably extending Definition 1.

Definition 1. An ordered system of pairwise disjoint recursive sets $[G_1, G_2, \ldots, G_h]$ of words over the alphabet Σ will be called an *information medium* (IM).

Definition 2. Let $\mathcal{A} = [\Sigma, S, \mathbf{M}_m{}^n, \pi_0, \alpha^1, \ldots, \alpha^h]$ be an FSA, $[G_1, \ldots, G_h]$ an IM, and δ, $0 \leqslant \delta \leqslant \frac{1}{2}$ a positive rational number. We shall say that $\mathcal{A}$ *represents* $[G_1, \ldots, G_h]$ *with parameter* δ if, for any word x over Σ,

$$x \in G_i \supset p_{\mathcal{A}}(i, x) = \pi_0 \mathbf{M}_x \alpha^i \nless \frac{1}{2} + \delta.$$

The terminology just introduced is also convenient in formulation of the following problems.

Suppose we are given certain IMs over Σ which can be represented by FDA (this is not always possible). Fixing the parameter δ, we consider the following problem: find conditions under which an IM over Σ is representable with parameter δ by an FSA $\mathcal{A} = [\Sigma, S, \mathbf{M}_m{}^n, \pi_0, \alpha^1, \ldots, \alpha^h]$ which has less internal states than any FDA representing the IM. The existence of such IMs for any δ was proved by Kovalenko (1965).* Rabin's example (Rabin, 1963), demonstrating how states are saved in FSA in comparison with FDA, is only partially relevant to this formulation of the problem. For in his example Rabin considered events $T(\mathcal{A}, \lambda)$ with $\lambda \neq \frac{1}{2}$, and states are saved by narrowing down the "region of isolation" of the cut-point.

At the present time a full solution of the problem is still quite far off. We have therefore seen fit to present below a certain class of examples, which throw light on the nature of the problem and the possible modes of approach to its solution.

Throughout the sequel, Σ will denote the two-letter alphabet $\{\sigma_1, \sigma_2\}$. For any word x over Σ, $l_1(x)$ and $l_2(x)$ will denote the number of occurrences of σ_1 and σ_2, respectively, in x. We now define an IM $[G_1^m, G_2^m]$ over Σ for every natural number m.

A word x will belong to G_1^m if and only if there are words u, v, w such that

$$\begin{gathered} x \equiv uvw; \\ l(x) = 4m; \\ l(u) + l(v) \leqslant 2m; \\ l(v) = l_2(v) = m; \\ l_1(uw) = l(uw). \end{gathered}$$

A word y will belong to G_2^m if and only if there are words u', v', w' such that

$$\begin{gathered} y \equiv u'v'w'; \\ l(y) = 4m; \\ l(v') + l(w') \leqslant 2m; \\ l(v') = l_2(v') = m; \\ l_1(u'w') = l(u'w'). \end{gathered}$$

Lemma 1. Any FDA $\mathcal{A} = [\Sigma, S, \mathbf{M}_2^n, \alpha^1, \alpha^2]$** *that represents the IM* $[G_1^m, G_2^m]$ *has at least* $2E(\sqrt{m})$ *internal states.*

Proof. It is clear that, given $\mathcal{A}$, we can determine an algorithm $\mathfrak{A}$, applicable to any word $s_i x$ over $S \cup \Sigma$ (where x is a nonempty word over Σ) and producing s_j if and only if x steers $\mathcal{A}$ from state s_i to state s_j with probability 1. Henceforth we shall omit the phrase "with probability 1" and, instead of "x steers $\mathcal{A}$ from state s_i to state s_j with probability 1," we shall write $\mathfrak{A}(s_i x) \equiv s_j$.

Let k denote the largest natural number such that

$$\begin{gathered} \forall x \forall y ((l(xy) = l_1(xy) \& l(x) \neq l(y) \& \\ \& \max(l(x), l(y)) < k) \supset \mathfrak{A}(s_1 x) \not\equiv \mathfrak{A}(s_1 y)). \end{gathered} \tag{31}$$

There are two possibilities: $k \geqslant m$ or $k < m$. In the first case, if $m > 1$,† we at once obtain the required estimate for the number of internal states

* We are here disregarding the fact that Kovalenko neglects to prove the important statement that the corresponding IMs cannot be represented by FDA if the latter have at most $n/3$ internal states.

** From now on we shall omit the vector π_0 in specification of an FDA, assuming that the initial state is s_1.

† It is obvious that $\mathcal{A}$ must have at least two states.

of $\mathcal{A}$. The second case requires further consideration. Let $t<k$ be a natural number such that, for any words x', $l(x')=l_1(x')=t$, and x'', $l(x'')=l_1(x'')=k-1$, we have $\mathfrak{A}(s_1x') \equiv \mathfrak{A}(s_1x''\sigma_1)$.

Now let h denote the smallest natural number such that $(k-1)+(k-t)h \geqslant 2m$, and h' the largest natural number such that $(k-1)+(k-t)h' \leqslant 3m$.

We claim that the number n of internal states of $\mathcal{A}$ satisfies the inequality

$$n \geqslant k+\min(h'-h+1,\ m-k+1). \tag{32}$$

To prove this, we let z be a word over Σ such that $l(z)=l_2(z)=m$.

We first show that $\mathfrak{A}(s_1x) \not\equiv \mathfrak{A}(s_1x''z)$ for any word over Σ such that $l(x)=l_1(x)<k$.

Suppose the contrary: there is a word x^*, $l(x^*)=l_1(x^*)<k$, such that $\mathfrak{A}(s_1x^*) \equiv \mathfrak{A}(s_1x''z)$. Then there is a word u, $l(u)=l_1(u)=(k-t)h$, such that $\mathfrak{A}(s_1x^*) \equiv \mathfrak{A}(s_1x''uz)$. We can now find words w_1, w_2 over Σ such that $x''zw_1w_2 \in G_1{}^m$ and $x''uzw_1 \in G_2{}^m$.

By our assumptions,

$$\begin{cases} \mathfrak{A}(s_1x''zw_1w_2) \equiv \mathfrak{A}(s_1x^*w_1w_2); \\ \mathfrak{A}(s_1x''uzw_1) \equiv \mathfrak{A}(s_1x^*w_1). \end{cases}$$

But since $l(w_1)=l_1(w_1) \geqslant t$ and $l(w_2)=l_1(w_2)=(k-t)h$, it follows that

$$\mathfrak{A}(s_1x^*w_1w_2) \equiv \mathfrak{A}(s_1x^*w_1).$$

Thus the automaton $\mathcal{A}$ cannot distinguish between the words $x''zw_1w_2$ and $x''uzw_1$. This contradiction proves our assertion.

For brevity, let us denote the internal state $\mathfrak{A}(s_1x''z)$ by s^*. It is easy to see that the internal states $\mathfrak{A}(s^*y)$ generated by the words y, $l(y)=l_1(y) \leqslant m-k+1$ are distinct from the internal states $\mathfrak{A}(s_1x)$ generated by the words x, $l(x)=l_1(x)<k$.

Indeed, since $l(w_1)=3m-(k-1)-(k-t)h$, we have $l(w_1)-(m-k)-t=2m-(k-t)h-t+1$. But

$$t-1+(k-t)h=(k-1)+(k-t)(h-1)<2m.$$

Consequently, $l(w_1)-(m-k)>t$, and so the internal states $\mathfrak{A}(s_1x)$ and $\mathfrak{A}(s^*y)$ generated by the appropriate words x, y over Σ are distinct.

We now let ν' be the largest natural number ν satisfying the condition

$$\begin{aligned} &\nu \leqslant m-k+1 \,\&\, \forall u \forall v ((l(uv)=l_1(uv) \,\&\, l(u) \neq l(v) \,\&\\ &\& \max(l(u),\ l(v)) \leqslant \nu) \supset \mathfrak{A}(s^*u) \not\equiv \mathfrak{A}(s^*v)). \end{aligned} \tag{33}$$

We claim that

$$\nu' \geqslant \min(h'-h+1,\ m-k+1). \tag{34}$$

Suppose the contrary:

$$\nu' < \min(h'-h+1,\ m-k+1),$$

and find a natural number $t' < \nu'$ such that for any words y', $l(y') = l_1(y') = t'$, and y'', $l(y'') = l_1(y'') = \nu'$,

$$\mathfrak{A}(s^* y') \cong \mathfrak{A}(s^* y'' \sigma_1).$$

By our assumption on ν',

$$\nu' - t' < h' - h + 1.$$

We denote the number $3m - (k-1) - (k-t)h'$ by μ and find a natural number $l \leqslant h' - h$ for which

$$\begin{cases} \mu + (k-t) l < t'; \\ \mu + (k-t)(l+1) \geqslant t'. \end{cases}$$

It is readily seen that $\nu' - t' \leqslant \nu' - l - 1 < h' - h - l$. Consequently, there exist natural numbers i, $i \leqslant \nu' - l - 1$, and j such that

$$h' - l - i - 1 = (\nu' - t') j. \tag{35}$$

Now let $\tilde{u}$ and $\tilde{w}$ be words such that $l(\tilde{u}) = l_1(\tilde{u}) = (\nu' - t') j (k-t)$ and $l(\tilde{w}) = l_1(\tilde{w}) = \mu + (l+i+1)(k-t)$. Then the word $\tilde{y} = x'' \tilde{u} z \tilde{w}$ belongs to $G_2{}^m$. But it is easy to see that the automaton $\mathcal{A}$ cannot distinguish the word $x'' z w_1 w_2$, which is in $G_1{}^m$, from $\tilde{y}$. This contradiction proves formula (34). But since $n \geqslant k + \nu'$, this also establishes formula (32).

Suppose now that

$$\min(h' - h + 1, m - k + 1) = h' - h + 1;$$

then

$$n \geqslant k + E\left(\frac{m}{k}\right). \tag{36}$$

We now find an integer i such that $k = E(\sqrt{m}) + i$ and write (36) as

$$n \geqslant E(\sqrt{m}) + i + E\left(\frac{m}{E(\sqrt{m}) + i}\right). \tag{37}$$

Since

$$(E(\sqrt{m}))^2 \leqslant m$$

we infer from (37) that

$$n \geqslant E\left(E(\sqrt{m}) + i + \frac{(E(\sqrt{m}))^2}{E(\sqrt{m}) + i}\right) = E\left(\frac{2(E(\sqrt{m}))^2 + 2iE(\sqrt{m}) + i^2}{E(\sqrt{m}) + i}\right) \geqslant$$
$$\geqslant E\left(2E(\sqrt{m}) \frac{E(\sqrt{m}) + i}{E(\sqrt{m}) + i}\right) = 2E(\sqrt{m}).$$

Consequently, if $\min(h' - h + 1, m - k + 1) = h' - h + 1$, then $n \geqslant 2E(\sqrt{m})$.

The same bound is obtained for n if $\min(h'-h+1, m-k+1)=m-k+1$. This means that if $k<m$ the desired estimate for the number of internal states of $\mathcal{A}$ is valid, and the proof of Lemma 1 is complete.

For the sequel, we shall find it convenient to assume that the FDA are equipped with an "output," i. e., the internal states of the FDA may be associated with certain letters of the alphabet Σ. This correspondence Φ is specified for any FDA $\mathcal{A}=[\Sigma, S, \mathbf{M}_2{}^n, \alpha^1, \alpha^2]$ by a condition of the type $(\Phi(s_i) \equiv \sigma_j) \leftrightarrow (\alpha_i{}^j=1)$, $j=1, 2$.

Let $\widetilde{\mathcal{A}}=[\Sigma, \widetilde{S}, \widetilde{\mathbf{M}}_2{}^3, \widetilde{\alpha}^1, \widetilde{\alpha}^2]$ be an FDA with

$$\widetilde{\mathbf{M}}_{\sigma_1}=\begin{pmatrix}0&1&0\\1&0&0\\0&0&1\end{pmatrix};\quad \widetilde{\mathbf{M}}_{\sigma_2}=\begin{pmatrix}0&0&1\\0&0&1\\0&0&1\end{pmatrix};\quad \widetilde{\alpha}^1=\begin{pmatrix}1\\0\\0\end{pmatrix}.$$

We define an IM $[\widetilde{G}_1{}^m, \widetilde{G}_2{}^m]$ over Σ as follows: $x\in\widetilde{G}_j{}^m$ if and only if $l(x)=4m$ and there exists $z\in G_j{}^m$ such that the k-th letter of x, $k=1, 2, \ldots, 4m$, is $\Phi(\mathfrak{A}(s_1 z_k))$, where z_k is the prefix of z of length k. It is clear that every word $x\in\widetilde{G}_1{}^m$ satisfies the condition $l_1(x)\leqslant\frac{1}{2}m$, and every word $y\in\widetilde{G}_2{}^m$ the condition $l_1(y)\geqslant m$.

Lemma 2. For any natural number m, one can construct an FSA $\mathcal{A}'=[\Sigma, S', \mathbf{M}_2'^n, \pi_0, \alpha'^1, \alpha'^2]$ with $n\leqslant 321^2$ internal states which represents the IM $[\widetilde{G}_1{}^m, \widetilde{G}_2{}^m]$ with parameter $\delta=\frac{1}{16}$.

Proof. If $m<320$, we can construct an FDA with $\frac{1}{2}m+1$ internal states which represents the IM $[\widetilde{G}_1{}^m, \widetilde{G}_2{}^m]$. Consequently, if $m<320$ we can construct an FSA $\mathcal{A}'$ satisfying the conditions of our lemma. If $m\geqslant320$, we define the components of $\mathcal{A}'$ as follows.

(a) The letters s'_i, $i=1, 2, \ldots, 321^2=n$, correspond to the ordered pairs $f\Box g$ of natural numbers such that $0\leqslant f, g\leqslant320$. The states s'_i are indexed in such a way that $f\Box g$ precedes $f'\Box g'$ if and only if $f<f'\vee(f=f'\&g<g')$.

(b) We define $\alpha_i'^1$ by the condition $(\alpha_i'^1=1)\leftrightarrow((s_i'\rightarrow f\Box g)\&(f<g))$.

(c) We shall first specify the matrices $\mathbf{M}'_{\sigma_1}$ and $\mathbf{M}'_{\sigma_2}$ only up to the values of parameters p and p', respectively. Stipulating that p and p' are positive rational numbers less than 1, we define $\mathbf{M}'_{\sigma_1}$ by

$$\forall i(s'_i\rightarrow f\Box g\supset((f<320\supset(a'_{ii}(\sigma_1)=1-p\,\& \\ \&\, a'_{i\,i+321}(\sigma_1)=p))\,\&\,(f=320\supset a'_{ii}(\sigma_1)=1))).$$

The matrix $\mathbf{M}'_{\sigma_2}$ is defined similarly:

$$\forall i(s'_i\rightarrow f\Box g\supset((g<320\supset(a'_{ii}(\sigma_2)=1-p'\,\& \\ \&\, a'_{i\,i+1}(\sigma_2)=p'))\,\&\,(g=320\supset a'_{ii}(\sigma_2)=1))).$$

(d) $\pi_0=(1, 0, \ldots, 0)$. It is easy to see that $\mathcal{A}'$ operates as a stochastic counter, counting the letters σ_1 and σ_2 independently.

We now observe that the matrix $\mathbf{M}'_{\sigma_2}$ is a direct sum of stochastic 321×321 matrices of the form

$$\begin{pmatrix} 1-p' & p' & 0 & 0 & \dots & \dots & 0 \\ 0 & 1-p' & p' & 0 & \dots & \dots & 0 \\ \cdot & \cdot & \cdot & \cdot & \cdot & \cdot & \cdot \\ 0 & \dots & \dots & \dots & 0 & 1-p' & p' \\ 0 & \dots & \dots & \dots & 0 & 0 & 1 \end{pmatrix}.$$

The number of terms in this sum is of course 321. It follows at once from the definition of $\mathbf{M}'_{\sigma_1}$ and the order imposed on the states that $\mathbf{M}'_{\sigma_1}$ is a product of three stochastic matrices: $\mathbf{M}'_{\sigma_1}=\mathbf{TAT}^{-1}$, where $\mathbf{T}$ (and hence also $\mathbf{T}^{-1}$) is a permutation matrix and $\mathbf{A}$ is a direct sum of stochastic 321×321 matrices of the form

$$\begin{pmatrix} 1-p & p & 0 & 0 & \dots & \dots & 0 \\ 0 & 1-p & p & 0 & \dots & \dots & 0 \\ \cdot & \cdot & \cdot & \cdot & \cdot & \cdot & \cdot \\ 0 & \dots & \dots & \dots & 0 & 1-p & p \\ 0 & \dots & \dots & \dots & 0 & 0 & 1 \end{pmatrix}.$$

In view of the above properties of the matrices $\mathbf{M}'_{\sigma_1}$, $\mathbf{M}'_{\sigma_2}$, we readily show that

$$\mathbf{M}'_{\sigma_1}\mathbf{M}'_{\sigma_2}=\mathbf{M}'_{\sigma_2}\mathbf{M}'_{\sigma_1}. \tag{38}$$

By (38), for any word x over Σ,

$$\pi_0\mathbf{M}'_x=\pi_0(\mathbf{M}'_{\sigma_1})^{l_1(x)}(\mathbf{M}'_{\sigma_2})^{l_2(x)}=\pi_0(\mathbf{M}'_{\sigma_2})^{l_2(x)}(\mathbf{M}'_{\sigma_1})^{l_1(x)}. \tag{39}$$

Suppose that y and z are words over Σ such that $l(y)=l_1(y)=l_1(x)$ and $l(z)=l_2(z)=l_2(x)$.

It is fairly easy to verify the following properties of the entries of the matrix $\mathbf{M}'_y$:

$$\left\{\begin{array}{l} a'_{i\ i+321h}(y)=C_{l(y)}{}^h p^h(1-p)^{l(y)-h}; \\ 0\leqslant h<\nu=\min\left(l(y),\ 321-\left]\dfrac{i}{321}\right[\right); \\ \sum\limits_{h=0}^{\nu} a'_{i\ i+321h}(y)=1. \end{array}\right. \tag{40}$$

The entries of $\mathbf{M}'_z$ possess similar properties:

$$\left\{\begin{array}{l} a'_{i\ i+\tilde{h}}(z)=C_{l(z)}{}^{\tilde{h}}(p')^{\tilde{h}}(1-p')^{l(z)-\tilde{h}}; \\ 0\leqslant\tilde{h}<\tilde{\nu}=\min\ (l(z),\ \mathrm{res}\ (i-1,\ 321)); \\ \sum\limits_{\tilde{h}=0}^{\tilde{\nu}} a'_{i\ i+\tilde{h}}(z)=1. \end{array}\right. \tag{41}$$

Denoting the i-th component of the stochastic vector $\pi_0\mathbf{M}'_x$ by $p_i(x)$, $i=1,\ 2,\ \dots,\ n$, and letting $s'_i\to f\square g$, we obtain

$$p_i(x) = a'_{1,\,321f+1}(y)\,a'_{321f+1,\,321f+g+1}(z) =$$
$$= a'_{1,\,g+1}(z)\,a'_{g+1,\,321f+g+1}(y) = a'_{1,\,321f+1}(y)\,a'_{1,\,g+1}(z)$$

(see formulas (39)—(41)).

Thus, for any i, if $s'_i \to f\Box g$,

$$p_i(x) = a'_{1,\,321f+1}(y)\,a'_{1,\,g+1}(z). \tag{42}$$

We now define a probability field $\Omega = (S', \mathfrak{B})$ by setting $\mathfrak{B}(s'_i) = p_i(x)$.

We define random variables $\mathfrak{P}_1$ and $\mathfrak{P}_2$ over Ω as follows: $\mathfrak{P}_1(s'_i) = f$; $\mathfrak{P}_2(s'_i) = g$ if $s'_i \to f\Box g$.

It is easy to see that $\mathfrak{P}_1$ and $\mathfrak{P}_2$ are independent random variables. (see (42)).

Let E denote the set of ordered pairs of natural numbers $f'\Box g'$ such that $0 \leqslant f' \leqslant l(y)$, $0 \leqslant g' \leqslant l(z)$. We define an algorithm $\mathfrak{B}'$ over E by

$$\mathfrak{B}'(f\Box g) = C_{l(y)}{}^{f'} p^{f'} (1-p)^{l(y)-f'} C_{l(z)}{}^{g'} (p')^{g'} (1-p')^{l(z)-g'}.$$

It is clear that $\Omega' = (E, \mathfrak{B}')$ is a probability field. Over this field we define random variables $\mathfrak{P}'_1$ and $\mathfrak{P}'_2$ by $\mathfrak{P}'_1(f'\Box g') = f'$, $\mathfrak{P}'_2(f'\Box g') = g'$. These random variables are independent.

Suppose now that $x \in \widetilde{G}_1{}^m$ and $M(\mathfrak{P}'_1) < M(\mathfrak{P}'_2)$. Then

$$M(\mathfrak{P}'_2) - M(\mathfrak{P}'_1) \geqslant \left(\frac{7}{2}p' - \frac{1}{2}p\right)m. \tag{43}$$

For $x \in \widetilde{G}_2{}^m$, the assumption $M(\mathfrak{P}'_2) < M(\mathfrak{P}'_1)$ yields

$$M(\mathfrak{P}'_1) - M(\mathfrak{P}'_2) \geqslant (p - 3p')\,m. \tag{44}$$

Letting c denote the number such that $cm = 80$, we define p and p' by the system of equations

$$\begin{cases} -\dfrac{1}{2}p + \dfrac{7}{2}p' = c; \\ p - 3p' = c. \end{cases}$$

This gives

$$\begin{cases} p = \dfrac{13}{4}c; \\ p' = \dfrac{3}{4}c. \end{cases}$$

The following properties of the random variables $\mathfrak{P}'_1$ and $\mathfrak{P}'_2$ are easily verified:

$$\begin{cases} \max\limits_{x\in\tilde{G}_1^m} M(\mathfrak{P}'_1) \leqslant \frac{13}{4} c \frac{1}{2} m = 130; \\ \max\limits_{x\in\tilde{G}_1^m} M(\mathfrak{P}'_2) = \frac{3}{4} c\cdot 4m = 240; \end{cases} \tag{45}$$

$$\begin{cases} \max\limits_{x\in\tilde{G}_2^m} M(\mathfrak{P}'_1) = \frac{13}{4} cE\left(\frac{3}{2}m\right) = 390 - \frac{13}{4} c\left(\frac{1}{2}m - E\left(\frac{1}{2}m\right)\right); \\ \max\limits_{x\in\tilde{G}_2^m} M(\mathfrak{P}'_2) = \frac{3}{4} c\cdot 3m = 180. \end{cases} \tag{46}$$

It follows from (40)–(42) and (45) that

$$\forall x\Big(x\in\tilde{G}_1^m \supset \Big(p\Big(\Big\{\mathfrak{P}'_1 > M(\mathfrak{P}'_1) + \frac{1}{2}cm - 1\Big\}\Big) < p\Big(\Big\{|\mathfrak{P}'_1 - M(\mathfrak{P}'_1)| > \\ > \frac{1}{2}cm - 1\Big\}\Big) \& p\Big(\Big\{\mathfrak{P}'_2 < M(\mathfrak{P}'_2) - \frac{1}{2}cm\Big\}\Big) < \\ < p\Big(\Big\{|\mathfrak{P}'_2 - M(\mathfrak{P}'_2)| > \frac{1}{2}cm\Big\}\Big)\Big)\Big).$$

Hence, by definition of α'^1 and formulas (43), (45), we get

$$\forall x\left(x\in\tilde{G}_1^m \supset p_{\mathcal{A}'}(1, x) > \left(1 - \frac{4M(\mathfrak{P}'_1)}{(cm)^2}\right)\left(1 - \frac{4M(\mathfrak{P}'_2)}{(cm)^2}\right) > \left(1 - \frac{1}{4}\right)^2\right). \tag{47}$$

Suppose now that x is in $\tilde{G}_2^m$, and let $\chi(x)$ denote the difference $M(\mathfrak{P}'_1) - M(\mathfrak{P}'_2)$

By (40)–(42) and (46),

$$\forall x\Big(x\in\tilde{G}_2^m \supset \Big(p\Big(\Big\{\mathfrak{P}'_1 < M(\mathfrak{P}'_1) - \frac{1}{2}\chi(x)\Big\}\Big)\Big(< p\Big(\Big\{|\mathfrak{P}'_1 - M(\mathfrak{P}'_1)| > \\ > \frac{1}{2}\chi(x)\Big\}\Big) \& p\Big(\Big\{\mathfrak{P}'_2 > M(\mathfrak{P}'_2) + \frac{1}{2}\chi(x) - 1\Big\}\Big) < \\ < p\Big(\Big\{|\mathfrak{P}'_2 - M(\mathfrak{P}'_2)| > \frac{1}{2}\chi(x) - 1\Big\}\Big)\Big)\Big).$$

Hence, by the definition of α'^2 (see also (44) and (46)), we get

$$\forall x\left(x\in\tilde{G}_2^m \supset p_{\mathcal{A}'}(2, x) > \left(1 - \frac{4M(\mathfrak{P}'_1)}{(cm)^2}\right)\left(1 - \frac{4M(\mathfrak{P}'_2)}{(cm)^2}\right) > \left(1 - \frac{1}{4}\right)^2\right). \tag{48}$$

Formulas (47) and (48) show that $\mathcal{A}'$ represents the IM $[\tilde{G}_1^m, \tilde{G}_2^m]$ with parameter $\delta = \frac{1}{16}$. By linking up the output of the FDA $\tilde{\mathcal{A}}$ to the input of the

FSA $\mathscr{A}'$, we obtain an automaton $\mathscr{A}''$ which represents the IM $[G_1^m, G_2^m]$ with parameter $\delta = \frac{1}{16}$. The distinctive feature of this automaton is that it provides a representation of the required IM in with delay 1. Since the number of internal states of $\mathscr{A}'$ is at most 321^2 for any m, it follows (see Lemma 1) that there exists a value of m such that any FDA representing the IM $[G_1^m, G_2^m]$ will have more internal states than $\mathscr{A}''$. By the normal representation theorem and Lemma 1, we conclude that such situations will arise for any value of the parameter δ', $0 < \delta' < \frac{1}{2}$. In other words, for any given δ' we can find m such that the IM $[G_1^m, G_2^m]$ is representable with parameter δ' by an FSA which has less internal states than any FDA representing the same IM.

We believe the IM $[G_1^m, G_2^m]$ is quite interesting. Unlike the IMs examined by Kovalenko (1965), $[G_1^m, G_2^m]$ consists of words x over Σ which are not distinguishable according to the number of occurrences of the letters σ_1 and σ_2, respectively.

NOTES

Paz (1966a) also gives an estimate for the number of internal states of an FDA representing an event $T(\mathscr{A}, \lambda)$ represented with isolated cut-point λ by an FSA $\mathscr{A}$, but our estimate is inferior.

According to our estimate, this number is at most $\left(1+\frac{n}{2\delta}\right)^n$ (where n is the number of internal states of $\mathscr{A}$), whereas Paz gives the estimate $\left(1+\frac{1}{2\delta}\right)^{n-1}$. It should not be thought, however, that this constructive estimate cannot be sharpened if so desired. The only reason we have refrained from doing this is to avoid obscuring the main idea of the constructive proof.

The normal representation theorem for IMs (special case: events) raises an important question: what is the minimal number of states of an FSA $\mathscr{A}'$ satisfying all the conditions of the theorem? Our constructive proof provides no estimate for this number, but it must be remembered that the general estimate we establish is far from satisfactory.

The problem of saving states, mentioned in passing by Ashby, was first considered in a rigorous and well-grounded manner by Rabin (1963) and later taken up by Kovalenko. The latter presented a theoretically important example showing what saving is to be expected in the most realistic situations. Kovalenko studied IMs $[G_1^m, G_2^m]$ defined as follows: G_1^m is the set of all words x over Σ of length m such that $l_1(x) \leqslant \frac{1}{3}m$; G_2^m is the set of all words x over Σ of length m such that $l_2(x) \leqslant \frac{1}{3}m$. As mentioned above, Kovalenko did not prove the estimate for the necessary number of internal states in an FDA representing the IM. However, the gist of his result remains valid for coarser estimates, since the main factor in this context is the rate of increase of the number of states. In personal correspondence with Kovalenko, we have suggested the bound $E\left(\sqrt{\frac{m}{3}}\right)$.

Since Kovalenko (1965) and Lorenz (1969b) consider only finite IMs, one is naturally interested in similar examples for infinite IMs (i. e., where G_1 and G_2 are infinite sets). Examples of this kind were studied by Metra (1971b). Somewhat narrowing down his definition, we might state that Metra considers IMs $[G_1, G_2]$ which satisfy the formula

$$\underset{x\in G_1}{\forall} x(l_2(x)\leqslant m)\ \&\ \underset{x\in G_2}{\forall} x(l_2(x)\geqslant m+h),$$

where m and h are fixed natural numbers. Here $l_2(x)$ may be replaced by $l_1(x)$. It turns out that, for sufficiently large m and h, a certain stochastic automaton, known as a stochastic counter, will represent IMs of this type with a relatively small number of internal states and sufficiently large radius of isolation. One can always find IMs of this type whose representation by FDA requires considerably more internal states than the corresponding FSA. Although Metra himself did not devote special attention to these special IMs, there is no doubt that they may indeed be constructed. The transition matrices of a stochastic counter have the form

$$\mathbf{M}_{\sigma_1}=\begin{pmatrix}1 & 0 & 0 & \dots & 0\\ 0 & 1 & 0 & \dots & 0\\ \dots & \dots & \dots & \dots & \dots\\ 0 & 0 & 0 & \dots & 1\end{pmatrix};\quad \mathbf{M}_{\sigma_2}=\begin{pmatrix}p & 1-p & 0 & 0 & \dots & 0\\ 0 & p & 1-p & 0 & \dots & 0\\ \dots & \dots & \dots & \dots & \dots & \dots\\ 0 & 0 & \dots & 0 & p & 1-p\\ 0 & 0 & \dots & 0 & 0 & 1\end{pmatrix}.$$

The column vector α is defined by

$$\alpha=\begin{pmatrix}0\\ 0\\ \dots\\ 0\\ 1\end{pmatrix}.$$

The present author has constructed another class of infinite (and also finite) IMs which dispel the doubts associated with the admissible nearness of the sets G_1 and G_2 when the FSA indeed reduces the number of states. The impression was that some saving could be achieved only provided there existed a natural number h such that

(1) $\forall x((l(x)\leqslant h)\supset(x\in G_1\cup G_2))$
(or, respectively $\forall x((l(x)\geqslant h)\supset(x\in G_1\cup G_2)))$;

(2) the sets $\{l(x)\leqslant h\}G_1$ and $\{l(x)\leqslant h\}G_2$ are not empty (or, respectively, the sets $\{l(x)\geqslant h\}G_1$ and $\{l(x)\geqslant h\}G_2$) are not empty). We have proved that for any radius of isolation δ and any natural number h there exist IMs satisfying these conditions whose representations by FSA yield an arbitrarily large saving of states.

Metra (1971a) brought these investigations to completion (in at least one theoretically important aspect) by considering complete IMs, i. e., IMs $[G_1, G_2]$ such that $\forall x(x \in G_1 \cup G_2)$.

He showed that, given a value of the parameter $\delta \leqslant \frac{1}{6}$, one can construct a sequence of IMs $[G_1^m, G_2^m]$ such that the m-th IM of the sequence may be represented by an FSA with parameter δ, using $6m$ internal states, while the appropriate FDA requires $2m(4m^2-1)$ states. Over a one-letter alphabet Σ, the IM $[G_1^m, G_2^m]$ is defined thus: G_1^m is the set of all words x over Σ whose length is a common multiple of at least two of the numbers $2m-1$, $2m$, $2m+1$.

In the context of abstract synthesis of FSA, one can also consider representation of the union (intersection) of two representable events, and representation of the complement of an event. Special cases of these problems have been solved (in a classical setting) by Bukharaev, but we are still far from a complete solution.

In the constructive approach, the following proposition may be proved:

There are representable events whose complements are not representable.

Proof. Let $\mathfrak{A}_d$ denote a normal algorithm over Σ such that

$$\forall x\left(\mathfrak{A}_d(x) = \frac{1}{2}d\right),$$

where $d \in \{B_g\}$. It is easy to see that every event $\{\mathfrak{A}_d(x) > 0\}$ is representable by the FSA $\mathcal{A}_d = [\Sigma, S, \mathbf{M}_1^2, \pi_0, \alpha]$, where

$$\mathbf{M}_{\sigma_1} = \begin{pmatrix} \frac{1}{2} - \frac{1}{2}d & \frac{1}{2} + \frac{1}{2}d \\ \frac{1}{2} - \frac{1}{2}d & \frac{1}{2} + \frac{1}{2}d \end{pmatrix}; \quad \pi_0 = (1, 0); \quad \alpha = \begin{pmatrix} 0 \\ 1 \end{pmatrix}.$$

The requisite cut-point here is $\lambda = \frac{1}{2}$. Suppose that the event $\{\mathfrak{A}_d(x) \not> 0\}$ is also representable. This means that for every d one can construct an FSA $\mathcal{A}'_d$ and a cut-point λ_d such that

$$\{\mathfrak{A}_d(x) \not> 0\} \approx T(\mathcal{A}', \lambda_d).$$

But this implies that the following formula is true:

$$\neg\left((p_{\mathcal{A}'_d}(\sigma_1) - \lambda_d > 0) \,\&\, \left(p_{\mathcal{A}_d}(\sigma_1) - \frac{1}{2} > 0\right)\right).$$

Now, by the principle of constructive choice,

$$\left((p_{\mathcal{A}'_d}(\sigma_1) - \lambda_d > 0) \vee \left(p_{\mathcal{A}_d}(\sigma_1) - \frac{1}{2} > 0\right)\right).$$

This is clearly equivalent to

$$((d > 0) \vee (d \not> 0)).$$

But since $d > 0$ implies $d \neq 0$ and $d \not> 0$ implies $d = 0$, we now have $((d=0) \vee (d \neq 0))$, contradicting the known properties of $\{B_g\}$. Thus the complements of the above events cannot be representable.

It should be clear that the greatest difficulties in solving these problems of abstract synthesis stem from the condition that the numerical parameters involved are rational numbers.

It would seem that outside the realm of rational numbers there is no point in attempting to develop methods for the construction of reduced or minimal FSA. This is a necessary condition for solution of those problems of abstract synthesis whose classical analogs were solved successfully by Carlyle, Bacon and Even (1965).

Chapter VIII

STRUCTURAL SYNTHESIS OF FINITE STOCHASTIC AUTOMATA

If finite stochastic automata were merely mathematical models for unreliable deterministic automata, there would be hardly any point in taking up the question of structural synthesis. In that case, the theory of FSA would be simply a tool for analysis of various negative phenomena inherent in actual FDA; the problems of structural synthesis would then be significant only for FDA or, alternatively, for FSA whose transition probabilities are close to 0 or to 1. This is by no means the case. FSA provide mathematical models for certain processes of stochastic control, statistical optimization and so on, all processes which are eminently worthy of simulation. Consequently, the range of problems of structural synthesis of FSA is by no means an imaginary entity, stimulated only by analogy with the theory of FDA; it arises quite naturally from the practical needs which the theory of FSA is intended to meet.

§1. BASIC CONCEPTS OF THE THEORY OF FINITE GRAPHS

The theoretical basis for all subsequent constructions is the theory of graphs, a few notions and definitions of which we now present. We follow mainly the classical books of Berge (1958), Ore (1962) and Harary et al. (1965).

Definition 1. Let E be a finite set of words over an alphabet B and $\mathfrak{F}$ a normal algorithm over the alphabet $B \cup \square$, applicable to any word $e \in E$. A system $G=[E, \mathfrak{F}]$ will be called a graph if, for any $e \in E$, either $\mathfrak{F}(e) \eqcirc \square$ or

$$\mathfrak{F}(e) \eqcirc \square e_1 \square \ldots \square e_k \square,$$

where e_i, $i=1, 2, \ldots, k$, are pairwise distinct elements of E, $k \leqslant \mu(E)$.

The elements of E are usually called the vertices of the graph.

Definition 2. Let $G=[E, \mathfrak{F}]$ be a graph and e_1, e_2 two of its vertices. We shall say that e_1 and e_2 are joined by an arc going from e_1 to e_2 if $\square e_2 \square \mho \mathfrak{F}(e_1)$; if $e_1 \eqcirc e_2$, we shall say elementary loop instead of arc. The symbolic notation for the fact that e_1 and e_2 are joined by an arc going from e_1 to e_2 will be $e_1 \ominus e_2$.

Definition 3. A path from a vertex e_1 to a vertex e_2 in a graph G is an ordered sequence C of vertices, $C=(e_{i_1}, e_{i_2}, \ldots, e_{i_k})$, such that all elements of the sequence (except possibly the first and last) are distinct and satisfy the condition

$$(e_1 \overline{\underline{\ominus}}\, e_{i_1})\&(e_{i_1} \ominus e_{i_2})\&\ldots\&(e_{i_{k-1}} \ominus e_{i_k})\&(e_{i_k} \overline{\underline{\ominus}}\, e_2).$$

If $e_1 \overline{\underline{\ominus}}\, e_2$, a path joining e_1 and e_2 will be called a cycle.

Definition 4. Two graphs $G_1=[E_1, \mathfrak{F}_1]$ and $G_2=[E_2, \mathfrak{F}_2]$ are said to be isomorphic if $\mu(E_1)=\mu(E_2)$ and there exists an algorithm $\mathfrak{B}$ such that

(1) $\forall x((x \in E_1) \supset (\mathfrak{B}(x) \in E_2))$;

(2) if e and e' are graphically distinct elements of E_1, then

$$\mathfrak{B}(e)\, \not\overline{\underline{\ominus}}\, \mathfrak{B}(e');$$

(3) for all e, e' in E_1,

$$(e \ominus e') \leftrightarrow (\mathfrak{B}(e) \ominus \mathfrak{B}(e')).$$

It is easy to see that graph isomorphism is a decidable relation.

Henceforth we shall assume for the sake of convenience that the alphabet B is $A=\{|\}$ and all sets will consist of nonempty words over this alphabet; more precisely: we shall specify E by conditions of the type

$$(x \in E) \leftrightarrow (1 \leqslant l(x) \leqslant n). \tag{1}$$

Since the only distinction between words over A is by length, we shall simply identify words of length n with the natural number n, so that our sets E will be defined by conditions of the type

$$(x \in E) \leftrightarrow (1 \leqslant x \leqslant n).$$

Identifying isomorphic graphs, we may (if necessary) do away with specification of a graph via the set E and algorithm $\mathfrak{F}$ (rigorously speaking, only this pair is "rightfully" a graph). It is frequently convenient to specify a graph by simply displaying the set E and providing a full list of all arcs, i. e., a list of ordered pairs $e_i \square e_j$ of elements of E satisfying the relation $e_i \ominus e_j$. Clearly, this determines the graph up to isomorphism. Hence one arrives quite easily at the customary visual, geometric specification of a graph as a system of circles (for the vertices) joined by directed lines (for the arcs). For example, the graph $G=[E, \mathfrak{F}]$ defined by the set $E \approx \{1 \leqslant x \leqslant 5\}$ and the normal algorithm with scheme

$$\begin{array}{r}
||||| \rightarrow \square\square \\
|||| \rightarrow \square\square \\
||| \rightarrow \cdot\, \square|||\square|||||\square \\
|| \rightarrow \cdot\, \square|||\square \\
| \rightarrow \cdot\, \square|\square||\square||||\square|||||\square \\
\square\square \rightarrow \cdot\, \square
\end{array}$$

may also be specified by the same set E and the list of arcs

$$C=[1\square 1,\ 1\square 2,\ 1\square 4,\ 1\square 5,\ 2\square 3,\ 3\square 4,\ 3\square 5].$$

The same graph is visually depicted as in Figure 1.

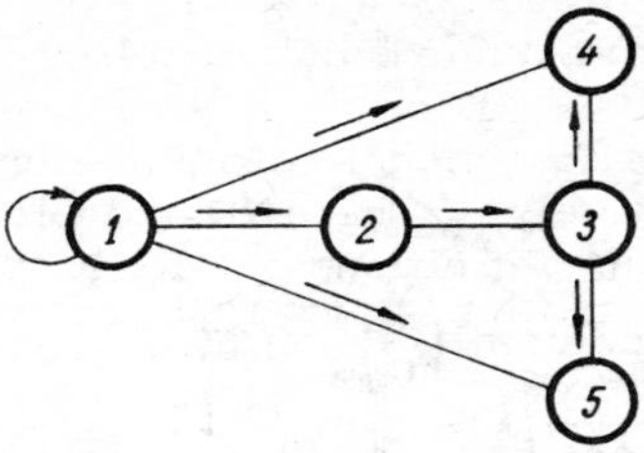

FIGURE 1.

R e m a r k. There are even interesting problems directly related to the visual representation of a graph, such as the problem of deciding what graphs possess a plane representation with nonintersecting arcs; such graphs are known as *planar graphs*.

§2. STRUCTURAL REALIZATION OF A STOCHASTIC VECTOR

For a rigorous formulation of the problem of structural synthesis of FSA, we now agree to confine attention (unless otherwise stipulated) to FSA with two output signals (0 and 1) and 2^m input signals ($\sigma_1, \sigma_2, \ldots, \sigma_{2^m}$). The latter may always be interpreted as words over the alphabet (0, 1) of length m, σ_i being the i-th word of length m in the lexicographic ordering. For example, if $m=2$, we interpret σ_1 as the word 00, σ_2 as 01, σ_3 as 10 and σ_4 as 11.

D e f i n i t i o n. *A finite basis* for synthesis of FSA (or *synthesis basis*) is a system of finite stochastic automata

$$\mathcal{B} =[\mathcal{A}_1, \mathcal{A}_2, \ldots, \mathcal{A}_c].$$

To eliminate possible misunderstandings arising from our definition, we add a few clarifications concerning the technological or physical interpretation of the basis. The latter is, or should be, a stock of mathematical models of certain technological or physical elements (components). When discussing a synthesis basis, one is usually imagining a certain stock of logic elements plus a memory element. For the engineer, a logic element is generally a device implementing the corresponding transformation of an incoming system of electrical pulses into an output pulse (including the absence of an output pulse); these are modeled mathematically by elementary boolean functions. For example, the logic function

$f(x, y) = (x\&y)$ is the mathematical model of an element with one output line and two input lines whose output line contains a pulse if and only if a pulse is applied in each input line simultaneously. This description of a logic element disregards the time delay present in any physical device, thus sometimes leading to too strong an idealization. Indeed, in certain cases the functioning of a real automaton cannot be understood without allowance for the time delay inherent in its components. One may well ask: may the basis for synthesis contain mathematical models for logic elements? The answer is in the affirmative. Below we describe a sequence of FSA $\mathcal{A}_\neg$, $\mathcal{A}_\vee$, $\mathcal{A}_\&$, which are mathematical models of logic elements (inverter, AND gate and OR gate, respectively). The AND and OR gates are allowed to have $m=2$ input lines. In our definitions, the letter β will denote the stochastic vector $(1, 0)$, α the column $\begin{pmatrix}1\\0\end{pmatrix}$ and α' the column vector $\begin{pmatrix}0\\1\end{pmatrix}$. $\mathbf{A}_0$ and $\mathbf{A}_1$ will denote the matrices $\begin{pmatrix}1&0\\1&0\end{pmatrix}$ and $\begin{pmatrix}0&1\\0&1\end{pmatrix}$, respectively. Finally, Σ_m and S_2 will denote the alphabets $\{\sigma_1, \sigma_2, \ldots, \sigma_{2^m}\}$ and $\{s_1, s_2\}$, respectively. With this notation:

$$\mathcal{A}_\neg = [\Sigma_1;\ S_2;\ \mathbf{A}_0;\ \mathbf{A}_1;\ \beta;\ \alpha];$$
$$\mathcal{A}_\vee = [\Sigma_2;\ S_2;\ \mathbf{M}_4^2;\ \beta;\ \alpha'], \text{where } \mathbf{M}_{\sigma_1} = \mathbf{A}_0,\ \mathbf{M}_{\sigma_2} = \mathbf{M}_{\sigma_3} = \mathbf{M}_{\sigma_4} = \mathbf{A}_1;$$
$$\mathcal{A}_\& = [\Sigma_2;\ S_2;\ \mathbf{M}_4^2;\ \beta;\ \alpha'],$$

where $\mathbf{M}_{\sigma_1} = \mathbf{M}_{\sigma_2} = \mathbf{M}_{\sigma_3} = \mathbf{A}_0$, $\mathbf{M}_{\sigma_4} = \mathbf{A}_1$.

A few words about time lags (or delays). As long as we were not concerned with structural synthesis, the time characteristics of the automata were immaterial. It would be quite legitimate to treat an FSA as a synchronous device, each of whose state changes takes place after a fixed time unit, or as an asynchronous device whose state changes occupy a time t that depends on the specific situation. We are now regarding the automata of the basis as synchronous devices with fixed time delay τ. By this token, we may view the FSA

$$\mathcal{A}_t = [\Sigma_1;\ S_2;\ \mathbf{A}_0;\ \mathbf{A}_1;\ \beta;\ \alpha']$$

with delay t as a mathematical model of a delay element. Finally, we note that the symbol 1 designates the presence of a pulse and 0 its absence. Although other interpretations are possible, we shall adhere to this one throughout the sequel.

Definition 6. Given a synthesis basis $\mathcal{B}$, we define a P-net with k_0 inputs and k_1 outputs as a system $L = [G, \mathfrak{h}]$, where G is a graph and $\mathfrak{h}$ a normal algorithm satisfying the following conditions:

(1) G contains no elementary loops;

(2) if ϱ is the number of vertices, then $\varrho \geqslant k_0 + k_1$;

(3) for each vertex n, $n \leqslant k_0$, in the graph G, the equation $x \ni n$ has no solution;

(4) if n is a vertex of G such that $k_0 < n < \varrho - k_1$, there exist vertices n_1 and n_2 $(n_1 \leqslant k_0) \& (\varrho - k_1 < n_2)$, such that G contains a path from n_1 to n and a path from n to n_2;

(5) for any vertex n, $\varrho-k_1<n$, G contains a vertex n', $n'\leqslant k_0$, such that n' and n may be joined by a path;

(6) the algorithm $\mathfrak{h}$ is applicable to any vertex n of G greater than k_0, producing an FSA of the basis $\mathcal{B}$; [we shall speak of this FSA as being attached to the vertex];

(7) if the equation $x \mathrel{\widehat{=}} n$ has exactly m solutions, the input alphabet of the FSA $\mathfrak{h}(n)$ is

$$\Sigma_m = (\sigma_1, \sigma_2, \ldots, \sigma_{2^m}).$$

Let us clarify the relation of the input states of the FSA attached to the graph vertices to the arcs which enter the vertices. Let n be a vertex of the graph G of a P-net $L=[G, \mathfrak{h}]$.

Suppose that $n_1, n_2, \ldots, n_c$ are the only vertices of G satisfying the equation $x \mathrel{\widehat{=}} n$. If

$$n_1<n_2<\ldots<n_c,$$

we shall call the arc $n_1\square n$ the first input line of the vertex, $n_2\square n$ the second, . . . , $n_c\square n$ the c-th. Suppose that at a given instant of time a signal α_i, $i=1, 2, \ldots, c$ is moving along the i-th input line of the vertex; then the input signal applied at that time to the FSA $\mathcal{A}$ attached to the vertex n is the signal σ_j corresponding to the word $\alpha_1\alpha_2\ldots\alpha_c$.

The outputs of the P-net are ordered in similar fashion. The first (second, third, etc.) output will be the first (second, third, etc.) when the outputs are arranged in increasing order as natural numbers. Here we shall define the i-th output of the P-net L to be the vertex $n=\varrho-k_1+i$. We illustrate this definition with a concrete example, using the synthesis basis

$$\mathcal{B}=[\mathcal{A}, \mathcal{A}_\iota, \mathcal{A}_\neg],$$

where $\mathcal{A}$ has input alphabet Σ_2, internal state alphabet $S=(s_1, s_2)$, transition matrices

$$\mathbf{M}_{\sigma_1} = (a_{ij});$$
$$\mathbf{M}_{\sigma_2} = (b_{ij});$$
$$\mathbf{M}_{\sigma_3} = (c_{ij});$$
$$\mathbf{M}_{\sigma_4} = (d_{ij}),$$

initial distribution $\pi_0=\beta$ and final state vector α'. A graphical depiction of a P-net over the above basis is shown in Figure 2.

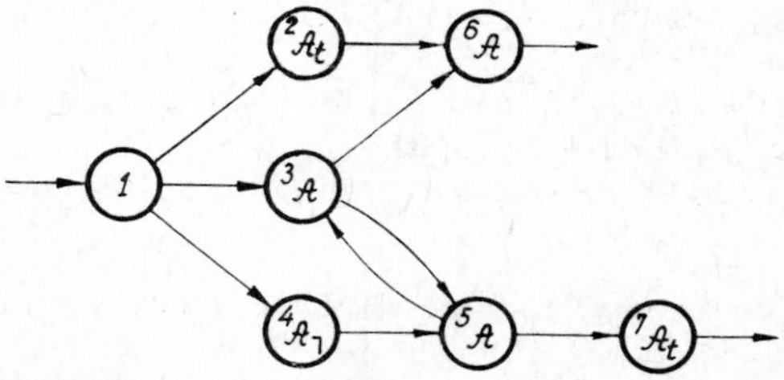

FIGURE 2.

With each P-net L we associate a set of words L_S — the *states* of L. Denote the internal state alphabet of the FSA $\mathfrak{h}(n)$, $n=k_0+1, k_0+2, \ldots, \varrho$, by $S_n=(s_{n_1}, s_{n_2}, \ldots, s_{n_c})$. We define L_S to be the set of all words of length $\varrho-k_0$ over the alphabet $S_{k_0+1} \cup S_{k_0+2} \cup \ldots \cup S_\varrho$, whose i-th letter is in S_{k_0+i}.

We shall call a state η of the P-net L an *initial state* if each letter of η is an initial state of the appropriate FSA $\mathfrak{h}(n)$. We now index all the elements in L_S from 1 to ν, denoting them by $\eta_1, \ldots, \eta_\nu$. This will be done in such a way that the first element is the initial state of L.

If L has k_0 input lines, it is natural in view of our interpretation to stipulate that the input alphabet of the P-net L contains exactly 2^{k_0} letters or 2^{k_0} trains of input signals. Denote the input letters of L by $\zeta_1, \ldots, \zeta_{2^{k_0}}$.

Next we define stochastic $\nu\times\nu$ matrices $\mathbf{L}_{\zeta_k}$, $k=1, 2, \ldots, 2^{k_0}$, letting the element $a_{ij}(\zeta_k)$ of $\mathbf{L}_{\zeta_k}$ be the probability that L goes from state η_i to η_j when the input to L is ζ_k. This number is readily evaluated, given the transition probabilities of the FSA $\mathfrak{h}(n)$. The formal definition is omitted here.

With each input word $g=\zeta_{s_1}\ldots\zeta_{s_d}$ of L we associate the stochastic matrix

$$\mathbf{L}_g=\mathbf{L}_{\zeta_{s_1}}\mathbf{L}_{\zeta_{s_2}}\ldots\mathbf{L}_{\zeta_{s_d}}.$$

If g is the empty word, $\mathbf{L}_g$ is defined to be the unit $\nu\times\nu$ matrix.

Letting π_g denote the stochastic vector which is the first row of $\mathbf{L}_g$, we now introduce the notion of a relatively attainable state.

A state η_i is said to be *attainable* relative to a system of input words $[e, g_1, \ldots, g_h]$ if there is a word

$$g \equiv e g_{t_1} g_{t_2} \ldots g_{t_l};$$

$$g_{t_m} \in \{(x \equiv g_1) \vee (x \equiv g_2) \vee \ldots \vee (x \equiv g_h)\},$$

such that the i-th component $p_i(g)$ of the vector π_g is positive.

We shall need yet another classification of the states of L: all the states which induce the same output word $\alpha \equiv \alpha_1\alpha_2\ldots\alpha_{k_1}$ will be collected into one class, which we shall simply call the class of states α.

Since the output words of a P-net L may also be viewed as boolean vectors, it is legitimate to use the notation ν_α for the number of occurrences of 1 in α. *

Definition 7. A P-net L is said to be *normal* relative to a system of input words $[e, g]$ if:

(1) no state of L in a class α for which $\nu_\alpha>1$ is attainable relative to the system $[e, g]$;

(2) There exists a sequence of finite sets of states of L, say $F_1, \ldots, F_{k_1+1}$, such that

(a) if η_s is a state in a class α, $\nu_\alpha=0$, which is attainable relative to $[e, g]$, then η_s is in the set F_{k_1+1};

(b) if η_s is a state in a class $\alpha=\alpha_1\alpha_2\ldots\alpha_{k_1}$, $\nu_\alpha=1$, $\alpha_i=1$, which is attainable relative to $[e, g]$, then η_s is in the set F_i;

(c) given a matrix $\mathbf{L}_g$ and any pair of sets F_i, F_j, there exists a number $a'_{ij}(g)$ such that

* In the theory of boolean functions the symbol ν_α is known as the norm of the vector α.

$$\forall s((\eta_s \in F_i) \supset (a_{st_1}(g) + \ldots + a_{st_r}(g) = a'_{ij}(g)),$$

where $t_1, \ldots, t_r$ are the indices of all states of L in F_j.

It is easy to see that the numbers $a'_{ij}(g)$ may be chosen arbitrarily if F_i is empty. We shall nevertheless stipulate that in that case the numbers $a'_{i1}(g), a'_{i2}(g), \ldots, a'_{ik_1+1}(g)$ form a stochastic vector.

One possibility is to set $a'_{ij}(g) = \frac{1}{k_1+1}$, $j = 1, 2, \ldots, k_1+1$, or $a'_{i1} = 1$ and $a'_{ij+1} = 0$.

Assuming now that the P-net L is normal relative to a system of input words $[e, g]$, we let $\mathbf{L}'_g$ denote the matrix whose (i, j)-th entry is $a'_{ij}(g)$. It is clear that $\mathbf{L}'_g$ is a stochastic $(k_1+1) \times (k_1+1)$ matrix. It is evidently uniquely defined only if each set F_i, $i = 1, 2, \ldots, k_1+1$, contains at least one state of L which is attainable relative to $[e, g]$.

Still assuming that L is normal, we define a (unique) stochastic vector $\pi'_e = (p'_1(e), \ldots, p'_{k_1+1}(e))$ by

$$p'_j(e) = a_{1t_1}(e) + \ldots + a_{1t_r}(e). *$$

Our assumption that L is normal enables us to define yet another class of importance for the sequel, using the vector π'_e and matrix $\mathbf{L}'_g$ just defined. Define

$$p_e = p'_1(e) + p'_2(e) + \ldots + p'_{k_1}(e).$$

Let $\mathbf{L}^g$ denote the column vector whose components are the numbers

$$a'_i(g) = a'_{i1}(g) + \ldots + a'_{ik_1}(g).$$

Now suppose that p_e and all the elements of $\mathbf{L}^g$ are positive. Then we can define a unique stochastic vector $\widehat{\pi}_e$ and a unique matrix $\widehat{\mathbf{L}}_g$ by

$$\widehat{\pi}_e = \left(\frac{p'_1(e)}{p_e}, \frac{p'_2(e)}{p_e}, \ldots, \frac{p'_{k_1}(e)}{p_e}\right);$$

$$\widehat{a}_{ij}(g) = \frac{a'_{ij}(g)}{a'_i(g)}.$$

Definition 8. Let π_0 be a stochastic vector and $\mathfrak{B}$ a synthesis basis. We shall say that π_0 is realizable in $\mathfrak{B}$ if there exist a P-net L over $\mathfrak{B}$ and a system of input words $[e, g]$ such that

(1) L is normal relative to $[e, g]$;

(2) p_e and the entries of $\mathbf{L}_g$ are all positive;

(3) $\pi_0 = \widehat{\pi}_e = (\widehat{a}_{i1}(g); \widehat{a}_{i2}(g), \ldots, \widehat{a}_{ik_1}(g))$, $i = 1, 2, \ldots, k_1+1$.

The realization of π_0 is said to be synchronous if $p_e = a'_i(g) = 1$, asynchronous otherwise.

Proposition 1. There is no finite synthesis basis $\mathfrak{B}$ over which one can construct a synchronous or asynchronous realization of every stochastic vector; in other words, for any finite basis $\mathfrak{B}$ there is a stochastic vector

* The numbers $t_1, t_2, \ldots, t_r$ run through the indices of all states in F_j.

π_0 *which is not realizable either synchronously or asynchronously in any P-net over this basis.*

This follows directly from the fact that the set of CRNs in the interval [0, 1] is nonenumerable.

Proposition 2. There is no finite synthesis basis $\mathcal{B}$ over which one can construct a synchronous realization for every stochastic vector with rational components.

This follows from the details of Definition 8 and the fact that the entries of the matrix $\widehat{\mathbf{L}}_g$ are computed from the transition probabilities of the basic FSA by operations of addition, multiplication and division by 1.

§3. COMPLETENESS OF THE BASIS

The first question we discuss here concerns the problem of completeness.

Definition 9. A synthesis basis $\mathcal{B}$ is said to be (weakly) complete if it admits synchronous realization of a set of stochastic vectors π_0 which is dense in the space of all stochastic vectors, i. e., for every stochastic vector π_0 and any number $\varepsilon=2^{-k}$ there is a vector π'_0, $\omega(\pi_0-\pi'_0)<\varepsilon$, which is synchronously realizable in $\mathcal{B}$.

The basis $\mathcal{B}$ is said to be strongly complete if, for every stochastic vector π_0 with rational components, one can synthesize a B-net over $\mathcal{B}$ which asynchronously realizes π_0.

Practically speaking, the more important notion is that of ordinary (weak) rather than strong completeness. Strong completeness is not without interest for the general theory of structural synthesis, however. We shall therefore consider both types of completeness, both here and later with regard to the wider problem of structural synthesis or, in our new terminology, structural realization of FSA.

Proposition 3. There exist finite synthesis bases $\mathcal{B}$ which are both weakly and strongly complete.

Proof. Consider the basis

$$\mathcal{B}=[\mathcal{A}_{\neg},\ \mathcal{A}_{\&},\ \mathcal{A}_t,\ \mathcal{A}_{\frac{1}{2}}],$$

where

$$\mathcal{A}_{\frac{1}{2}}=[\Sigma,\ S,\ \mathbf{M}_2^2,\ \pi_0,\ \alpha].$$

Here

$$\mathbf{M}_{\sigma_1}=\begin{pmatrix}1&0\\1&0\end{pmatrix};\quad \mathbf{M}_{\sigma_2}=\begin{pmatrix}\frac{1}{2}&\frac{1}{2}\\ \frac{1}{2}&\frac{1}{2}\end{pmatrix};\quad \pi_0=(1,0);\quad \alpha=\begin{pmatrix}0\\1\end{pmatrix}.$$

Let $\pi_0=(p_1,\ p_2,\ \ldots,\ p_l)$ be a stochastic vector and ε a rational number. Find a stochastic vector π'_0 with rational components

$$p'_i=\frac{c_i}{2^n},\quad i=1,2,\ldots,l,$$

such that $\omega(\pi_0 - \pi'_0) < \varepsilon$, where c_i are natural numbers. Construct networks* $L_1, L_2, \ldots, L_l$ realizing boolean functions

$$f_1(x_1, x_2, \ldots, x_n),\ f_2(x_1, x_2, \ldots, x_n),\ \ldots,\ f_l(x_1, x_2, \ldots, x_n),$$

which take the values of 1 respectively for $c_1, c_2, \ldots, c_l$ sequences of values for the variables $x_1, x_2, \ldots, x_n$. We also stipulate that

$$\forall i \forall j((i \neq j) \supset (f_i(\alpha_1, \alpha_2, \ldots, \alpha_n) f_j(\alpha_1, \alpha_2, \ldots, \alpha_n) = 0)).$$

The P-net to be synthesized is illustrated schematically in Figure 3. It is assumed here that the length of path from input to output is the same for all the networks L_i. If this length is h, the P-net realizes π'_0 synchronously. This is readily verified by considering a system of input words $[e, g]$, in which e and g are graphically equal to the word $\overbrace{100\ldots0}^{h \text{ times}}$.

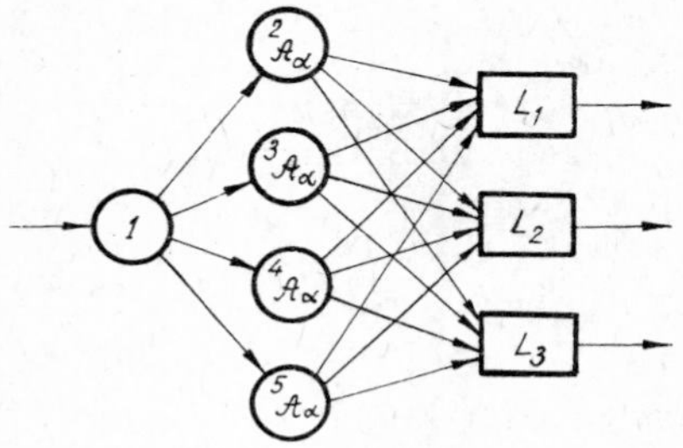

FIGURE 3.

That this basis is strongly complete was essentially proved by Makarevich and Giorgadze (1968). We say "essentially," since they adopt a different set of initial assumptions: their notion of a basis differs, albeit slightly, from ours; their P-nets are not defined rigorously and the definition of realization of a stochastic vector is rather inadequate. Last but not least, the statement of the problem and the method used are classical in spirit.

*Proposition 4. There is no algorithm that decides whether or not a synthesis basis is complete.***

Proof. We consider bases of the type

$$\mathcal{B} = [\mathcal{A}_{\neg}, \mathcal{A}_{\&}, \mathcal{A}_{\vee}, \mathcal{A}_t, \mathcal{A}_d],$$

where $\mathcal{A}_d = [\Sigma, S, \mathbf{M}_2^2, \pi_0, \alpha]$ is defined by

$$\mathbf{M}_{\sigma_1} = \begin{pmatrix} 1 & 0 \\ 1 & 0 \end{pmatrix};\quad \mathbf{M}_{\sigma_2} = \begin{pmatrix} 1-d & d \\ 1-d & d \end{pmatrix};\quad \pi_0 = (1, 0);\ \alpha = \begin{pmatrix} 0 \\ 1 \end{pmatrix}.$$

* Strictly speaking, P-nets over the basis $\mathcal{B}' = [\mathcal{A}_{\neg}, \mathcal{A}_{\&}, \mathcal{A}_t]$.

** That no such algorithm exists for strong completeness will follow from Proposition 5.

We claim that $\mathcal{B}$ is weakly complete if and only if $d>0$. Indeed, if $d=0$, the basis $\mathcal{B}$ consists solely of deterministic elements or FDA. Consequently, if $\mathcal{B}$ is weakly complete, then $d\neq 0$. Thus necessarily $d>0$.

Now let $d>0$: we shall show that then $\mathcal{B}$ is weakly complete.

To prove this, we use a construction analogous to that in the proof of Proposition 3. The structure of a P-net L realizing an ε-approximation of a stochastic vector π_0 is shown in Figure 3. The P-nets $L_1, L_2, \ldots, L_{k_1}$ are deterministic, since they are built up from the elements $\mathcal{A}_{\neg}$, $\mathcal{A}_{\&}$, $\mathcal{A}_{\vee}$ and $\mathcal{A}_t$. They realize certain boolean functions (to be defined below)

$$f_1(x_1, x_2, \ldots, x_n),\ f_2(x_1, x_2, \ldots, x_n),\ \ldots,\ f_{k_1}(x_1, x_2, \ldots, x_n),$$

i. e., they produce an output signal $\alpha=1$ if and only if their inputs receive signals $(\alpha_1, \alpha_2, \ldots, \alpha_n)$ such that

$$f_i(\alpha_1, \ldots, \alpha_n)=1, \quad 1\leqslant i\leqslant k_1.$$

The networks L_i are assumed to have identical time lags; in other words, the signal 1 is produced at the output of L_i after h time units τ, provided the input of L_i receives a signal sequence $(\alpha_1, \alpha_2, \ldots, \alpha_n)$ such that $f_i(\alpha_1, \ldots, \alpha_n)=1$. If $f_i(\alpha_1, \ldots, \alpha_n)=0$, the signal 0 will be produced at the output of L_i after h time units τ. It is also assumed that

$$\forall i \forall j((i\neq j)\supset(f_i(x_1, \ldots, x_n) f_j(x_1, \ldots, x_n)=0)$$

and

$$f_1(x_1, \ldots, x_n)\cup f_2(x_1, \ldots, x_n)\cup\ldots\cup f_{k_1}(x_1, \ldots, x_n)=1.$$

The number n of arguments is determined by the condition

$$\max(d^n, (1-d)^n)\not> \frac{\varepsilon}{3k_1}\not> \max(d^{n-j}, (1-d)^{n-j}). \tag{1}$$

It is desirable, though not essential, to have $j=1$.

The truth domains A_t of the functions $f_t(x_1, x_2, \ldots, x_n)$ are defined as follows.

For each sequence $\alpha=(\alpha_1, \ldots, \alpha_n)$ we define a number p_α by

$$p_\alpha=(1-d)^{1-\alpha_1} d^{\alpha_1} \ldots (1-d)^{1-\alpha_n} d^{\alpha_n}.$$

Index the numbers p_α from 1 to 2^n and denote them by $r_1, r_2, \ldots, r_{2^n}$. Now define numbers c_{1i}, $i=1, 2, \ldots, 2^n$, by $c_{1i}=r_1+r_2+\ldots+r_i$. Let ϱ_t be a rational number approximating p_t with maximum error $\frac{\varepsilon}{3k_1}$, and ϱ_{1i} a rational approximation to c_{1i} with the same maximum error. Choose a number ϱ_1 and find the first term in the sequence ϱ_{1i} such that

$$|\varrho_1-\varrho_{1i}|\leqslant\frac{\varepsilon}{3k_1}. \tag{2}$$

Suppose that (2) holds for $i=i_1$.

If $i_1 < 2^n$, define c_{2i} by

$$c_{2i} = r_{i_1+1} + r_{i_1+2} + \ldots + r_{i_1+i}.$$

Let ϱ_{2i} be a rational approximation of c_{2i} with maximum error $\frac{\varepsilon}{3k_1}$. Find the first term in the sequence ϱ_{2i} such that

$$|\varrho_2 - \varrho_{2i}| < \frac{\varepsilon}{3k_1}. \tag{3}$$

Suppose that (3) holds for $i = i_2$. If $i_1 + i_2 < 2^n$, a similar procedure applied to the sequence $(r_{i_1+i_2+1}, r_{i_1+i_2+2}, \ldots, r_{2^n})$ defines a number i_3. Assume, then, that we have already constructed the sequence $(i_1, i_2, \ldots, i_{k_1-1})$. The term i_t of this sequence is defined to be zero if $i_1 + i_2 + \ldots + i_{t-1} = 2^n$, and $i_{k_1} = 2^n - i_1 - i_2 - \ldots - i_{k_1-1}$.

Now define a sequence $(p'_1, p'_2, \ldots, p'_{k_1})$ by

$$p'_1 = r_1 + r_2 + \ldots + r_{i_1};$$

$$p'_2 = r_{i_1+1} + r_{i_1+2} + \ldots + r_{i_2+i_1};$$

$$\ldots\ldots\ldots\ldots\ldots\ldots$$

$$p'_{k_1} = r_{2^n} + r_{2^n-1} + \ldots + r_{2^n - i_{k_1}+1}.$$

Set $p'_t = 0$ if $i_t = 0$. It follows at once from the construction of the numbers p'_t, $t = 1, 2, \ldots, k_1 - 1$, that

$$|p_t - p'_t| < \frac{\varepsilon}{k_1}. \tag{4}$$

We claim that

$$|p_{k_1} - p'_{k_1}| < \varepsilon. \tag{5}$$

Indeed,

$$|p_{k_1} - p'_{k_1}| = |(1 - p_1 - p_2 - \ldots - p_{k_1-1}) - (1 - p'_1 - p'_2 - \ldots - p'_{k_1-1})| = |(p'_1 - p_1) + \ldots + (p'_{k_1-1} - p_{k_1-1})| \ngtr |p'_1 - p_1| + \ldots + |p'_{k_1-1} - p_{k_1-1}| \ngtr (k_1 - 1)\frac{\varepsilon}{k_1} < \varepsilon.$$

Thus $|p_t - p'_t| < \varepsilon$ for all t, $t = 1, 2, \ldots, k_1$.

Since

$$p'_1 + p'_2 + \ldots + p'_{k_1} = 1,$$

we have thus constructed a stochastic vector π'_0 such that $\omega(\pi_0 - \pi'_0) < \varepsilon$. But each number p'_t is the sum of some of the numbers r_i. The latter in turn are uniquely determined by the appropriate sequences $\alpha = (\alpha_1, \alpha_2, \ldots, \alpha_n)$.

Consequently, for each π_t we can define a set A_t of sequences α such that $\sum_{\alpha \in A_t} p_\alpha = p'_t$.

We may thus use the set A_t as the truth domain of the function $f_t(x_1, x_2, \ldots, x_n)$.

Of course, when the rules by which we constructed A_t are strictly observed it may turn out that some of these sets are empty. In such cases we need not consider the corresponding functions $f_t(x_1, x_2, \ldots, x_n)$ at all (if it is known in advance that $p_t=0$), i. e., the vector π_0 will have fewer components. There is an alternative, however: to construct a network realization for any identically false function $f_t(x_1, x_2, \ldots, x_n)$. Finally, it is also possible to modify the construction of the numbers i_t (and accordingly p'_t) in such a way that none of them vanish.

We now show that the resulting P-net L is normal relative to the system of input words $[e, g]$, where $e \trianglerighteq g \trianglerighteq \sigma_2 \overbrace{\sigma_1\sigma_1 \ldots \sigma_1}^{h \text{ times}}$.

Indeed, the states of L that produce an output word $\alpha \trianglerighteq (\nu_\alpha=0) \vee (\nu_\alpha>1)$ are not attainable relative to $[e, g]$. At the same time, any class of states α, i. e., any class of states producing an output word α, with $\nu_\alpha=1$, is attainable relative to $[e, g]$. The structure of the matrix $\mathbf{L}_g$ is easily deduced from the properties of the basis. The important point here is that the input σ_1 takes $\mathcal{A}_d$ into a definite state, independently of its previous state. The other elements of the basis $\mathcal{B}_d$ are deterministic, and so any input uniquely determines the next state.

Thus all rows of the matrix $\mathbf{L}_g$ are equal. Hence, by (2) and (3), we see that L is normal relative to $[e, g]$ and realizes the stochastic vector $\pi'_0=\widehat{\pi}_e=(\widehat{p}_1(e), \ldots, \widehat{p}_{k_1}(e))$, where $p_t(e) = \sum_{\alpha \in A_t} p_\alpha$. Consequently, $\omega(\pi_0-\widehat{\pi}_e)<\varepsilon$.

We have thus shown that the basis $\mathcal{B}_d$ is weakly complete if and only if $d>0$. But since there is no algorithm producing precisely the set of CRNs d, $0 \ntriangleright d \ntriangleright 1$, for which $d>0$, there can be no algorithm for the set of complete synthesis bases.

Proposition 5. The synthesis basis $\mathcal{B}=[\mathcal{A}_\neg, \mathcal{A}_\&, \mathcal{A}_\vee, \mathcal{A}_t, \mathcal{A}_d]$ is strongly complete if and only if $d>0$.

Proof. The arguments used in the proof of Proposition 4 imply that if $\mathcal{B}$ is strongly complete then $d>0$. Let us assume, then, that $d>0$ and show that $\mathcal{B}$ is strongly complete.

Let $\pi_0=(p_1, p_2, \ldots, p_l)$ be a stochastic vector with rational components. We wish to construct a P-net over $\mathcal{B}$, asynchronously realizing π_0.

Since π_0 is rational, we can find natural numbers $n, c_1, c_2, \ldots, c_l$, such that

$$p_1=\frac{c_1}{n};\ p_2=\frac{c_2}{n}, \ldots, p_l=\frac{c_l}{n}.$$

Now let $\mu=2m$ be a natural number such that $C_\mu^m \geqslant n$.

Our P-net will contain exactly μ vertices associated with the element $\mathcal{A}_d$, and one input line. Figure 3 illustrates the situation for $\mu=4$ and $l=3$.

The structure of our P-net in the general case is analogous; the principles governing its design are as follows. We first construct a P-net with $\mu+1$ vertices whose underlying graph contains the arcs $1\square 2, 1\square 3, \ldots, 1\square\mu$.

The vertices 2, 3, ..., μ are associated with $\mathcal{A}_d$. This net L has 2^μ states. Since the first state η_1 is the word $\overbrace{s_1 s_1 \ldots s_1}^{\mu \text{ times}}$, the matrix $\mathbf{L}_0$ has the form

$$\mathbf{L}_0 = \begin{pmatrix} 1 & 0 & 0 & \ldots & 0 \\ 1 & 0 & 0 & \ldots & 0 \\ \cdot & \cdot & \cdot & \cdot & \cdot \\ 1 & 0 & 0 & \ldots & 0 \end{pmatrix}.$$

The rows of $\mathbf{L}_1$ are also all equal; they are defined by the formula

$$\forall i \forall j ((\eta_j \cong s_{j_1} s_{j_2} \ldots s_{j_\mu}) \supset (a_{ij}(1) = (1-d)^{2-j_1} d^{|1-j_1|} \times \\ \times (1-d)^{2-j_2} d^{|1-j_2|} \ldots (1-d)^{2-j_\mu} d^{|1-j_\mu|})).$$

It is easy to see that the number of states of L containing exactly m occurrences of the letter s_1 (or, respectively, s_2) is just $C_\mu{}^m$. If η_j is a state of this type, then

$$a_{ij}(1) = (1-d)^m d^m. \qquad (6)$$

We now define $l+1$ finite sets of states of L by letting F_i, $i=1, 2, \ldots, l$, consist of the c_i states of L that contain exactly m occurrences of s_1, stipulating that $F_i \cap F_j \approx \varnothing$ when $i \neq j$.

The set F_{l+1} is defined by stipulating that $\eta_j \in F_{l+1}$ if j is not in $(F_1 \cup F_2 \cup \ldots \ldots \cup F_l)$. By the properties of the matrix $\mathbf{L}_1$, F_i is a set of equivalent states of the P-net L. We may thus group together states belonging to the same classes F_i, and when the corresponding procedure is applied to the matrix $\mathbf{L}_1$ the result is the matrix

$$\mathbf{B}_1 = \begin{pmatrix} b_{11}(1) & b_{12}(1) & \ldots & b_{1l}(1) & b_{1,l+1}(1) \\ b_{21}(1) & b_{22}(1) & \ldots & b_{2l}(1) & b_{2,l+1}(1) \\ \cdot & \cdot & \cdot & \cdot & \cdot \\ b_{l+1,1}(1) & b_{l+1,2}(1) & \ldots & b_{l+1,l}(1) & b_{l+1,l+1}(1) \end{pmatrix},$$

where

$$b_{1i}(1) = b_{2i}(1) = \ldots = b_{l+1,i}(1) = c_i(1-d)^m d^m.$$

[We shall call this procedure *condensing* the matrix with respect to the classes F_i.] Since $\sum_{i=1}^{l} c_i = n$, we have

$$\forall j \left(\frac{b_{ji}(1)}{\sum_{i=1}^{l} b_{ji}(1)} = \frac{c_i(1-d)^m d^m}{n(1-d)^m d^m} = \frac{c_i}{n} = p_i \right), \quad i=1, 2, \ldots, l.$$

Now construct combinational networks $L_1, L_2, \ldots, L_l$ with μ inputs and one output line. Suppose that L_j, $j=1, 2, \ldots, l$, realizes a boolean function

$f_j(x_1, x_2, \ldots, x_\mu)$, i. e., a signal 1 is produced at the output of L_j, after a certain time lag, if and only if the input receives a sequence of signals $(\alpha_1, \alpha_2, \ldots, \alpha_\mu)$ for which $f_j(\alpha_1, \alpha_2, \ldots, \alpha_\mu)=1$.

We further stipulate that the networks L_j have the same depth, i. e., the number of arcs in the shortest path from an input vertex to the output is a number h independent of L_j. We now construct a new P-net L^*, using the P-nets $L, L_1, L_2, \ldots, L_l$. The procedure is as follows. Identify the i-th input vertex $(i=1, 2, \ldots, \mu)$ of each P-net L_j with the $(i+1)$-th vertex of L. Let us analyze the structure of the matrix $\mathbf{L}^*_1$. We see that each entry $a^*_{ij}(1)$ of $\mathbf{L}^*_1$ is either zero or coincides with some entry of $\mathbf{L}_1$. For a more precise description we need the following notation: η_i will denote the i-th state of L, η_{jk} the k-th state of L_j, $\eta^*_t=\eta_{t_0}\eta_{1t_1}\eta_{2t_2}\ldots\eta_{lt_l}$ the t-th state of L^*. It is easy to see that for any given $\eta_{k_0}, \eta_{1k_1}, \eta_{2k_2}, \ldots, \eta_{lk_l}$ there is a sequence $\eta_{1u_1}, \eta_{2u_2}, \ldots, \eta_{lu_l}$ of states of L_j such that, for any η_{u_0},

$$a^*_{ku}(1)=p(\eta^*_k, \eta^*_u)=a_{k_0u_0}(1),$$

where

$$\eta^*_k=\eta_{k_0}\eta_{1k_1}\eta_{2k_2}\ldots\eta_{lk_l};$$

$$\eta^*_u=\eta_{u_0}\eta_{1u_1}\eta_{2u_2}\ldots\eta_{lu_l}.$$

Consequently, each row of the matrix $\mathbf{L}^*_1$ contains exactly 2^μ nonzero entries. Another important property of $\mathbf{L}^*_1$ is that, for any sequence of states $\eta^*_{t_1}, \eta^*_{t_2}, \ldots, \eta^*_{t_n}$ such that

$$a^*_{t_1t_2}(1)\cdot a^*_{t_2t_3}(1)\ldots a^*_{t_{d-1}t_d}(1)>0, \tag{7}$$

the state $\eta^*_{t_n}$ corresponds to an output of the P-net L^* which is completely determined by the first component of the state $\eta^*_{t_1}$. In other words, if the first component of $\eta^*_{t_1}$ is a state η_i corresponding to an output $(\alpha_{i_1}, \alpha_{i_2}, \ldots, \alpha_{i_\mu})$ of the P-net L, then the output of L^* when its state is $\eta^*_{t_n}$ is the word $\alpha^*_1\alpha^*_2\ldots\alpha^*_l$, where

$$\alpha^*_j=f_j(\alpha_{i_1}, \alpha_{i_2}, \ldots, \alpha_{i_\mu}).$$

We now assume that the truth domains A_j of the functions $f_j(x_1, x_2, \ldots, x_\mu)$ are disjoint. Then, in view of the above two properties of $\mathbf{L}^*_1$, one readily verifies that the only possible outputs of L^* at times $t=h+1, 2(h+1), 3(h+1), \ldots$ are words α with $\nu_\alpha\leqslant 1$. Thus the states of L^* attainable relative to $[e, g]$, $e\cong g=\overbrace{11\ldots 1}^{h+1\ \text{times}}$, may be grouped into $l+1$ classes $F^*_1, F^*_2, \ldots, F^*_{l+1}$, according as $\alpha_1, \alpha_2, \ldots,$ or α_l, is a one, or all are zero, when L^* is in state η^*_u. Again by the properties of $\mathbf{L}^*_1$, we have that the matrix $(\mathbf{L}^*_1)^{h+1}$ may be condensed with respect to our classification of the attainable states of L^* into classes $F^*_1, F^*_2, \ldots, F^*_{l+1}$; the entries b^*_{ij} of the condensed matrix $(\mathbf{L}^*_1)^{h+1}$ are defined by

$$b^*_{ij}=p(F^*_i, F^*_j)=\sum_{u\in U_j} a_{iu}(1),$$

where U_j is the set of indices of η_t for which the corresponding output of L is in A_j. It is now easy to see that the truth domains A_j may be so chosen that the matrices $\mathbf{B}_1$ and $\mathbf{B}^*=(b^*_{ij})$ coincide. Thus the P-net L^* asynchronously realizes π_0 relative to the above system $[e, g]$, and the proof of Proposition 5 is complete.

§4. STRUCTURAL REALIZATION OF A FINITE STOCHASTIC AUTOMATON

In this section we shall extend the problem of structural synthesis to FSA as a whole, rather than stochastic vectors. Definitions 7 and 8 will turn out to be special cases of the new definitions; however, as we shall see, these special cases are in fact the most important, and we shall essentially continue to work with them.

Let L be a P-net with k_1 output lines and system of input words $[e, g_1, g_2, \ldots, g_h]$. Let $s_1, s_2, \ldots, s_c$ be all the indices of states of L in the set F_k, and $t_1, t_2, \ldots, t_d$ the indices of the states in F_l. We define a vector $\pi_{kl}^{g_\nu}=(b_1, b_2, \ldots, b_c)$ whose components are entries of the matrix $\mathbf{L}_{g_\nu}$:

$$b_i=a_{s_it_1}(g_\nu)+a_{s_it_2}(g_\nu)+\ldots+a_{s_it_d}(g_\nu). \tag{8}$$

The components of the vector $\pi_{kl}^{g_\nu}$ are defined to be zero if F_l is empty, and undefined if F_k is empty.

Definition 10. We shall say that L is *normal* relative to $[e, g_1, \ldots, g_h]$ if the following conditions are satisfied.

(1) No state η corresponding to an output word α with $\nu_\alpha>1$ is attainable relative to $[e, g_1, \ldots, g_h]$.

(2) There is a sequence of finite sets $F_1, F_2, \ldots, F_{k_1}, F_{k_1+1}, \ldots, F_{2k_1}$, such that

(a) if η is an attainable state relative to $[e, g_1, \ldots, g_h]$ and the corresponding output word α has 1 at the i-th position (counting from the beginning), then η is in F_i;

(b) if $i\neq j, 1\leqslant i, j\leqslant k_1$, then

$$[F_i\cap F_j\approx\varnothing]\&[F_{k_1+i}\cap F_{k_1+j}\approx\varnothing];$$

(c) $F_{k_1+1}\cup F_{k_1+2}\cup\ldots\cup F_{2k_1}$ is a subset of the set of states η for which the corresponding output word α as $\nu_\alpha=0$;

(d) if η is attainable relative to $[e, g_1, \ldots, g_h]$ and the corresponding output word α has $\nu_\alpha=0$, then η is in the set

$$F_{k_1+1}\cup F_{k_1+2}\cup\ldots\cup F_{2k_1};$$

(e) if $(s\neq t)\&(\eta_i\in F_s)\&(\eta_j\in F_{k_1+t})$, then $a_{ij}(g_\nu)=0$;

(f) $a_{ij}(g_\nu)$ also vanishes when $(\eta_i\in F_{k_1+s})\&(\eta_j\in F_{k_1+t})\&(s\neq t)$.

(3) For each g_ν, $\nu=1, 2, \ldots, h$, and each pair of sets (F_k, F_l), all the components of the vector $\pi_{kl}^{g_\nu}$ are equal.

Let $a'_{kl}(g_\nu)$ denote the first component of the vector $\pi_{kl}{}^{g_\nu}$, provided of course that the vector is defined (i. e., F_k is not empty). In view of the sequence of finite sets $F_1, F_2, \ldots, F_{2k_1}$ figuring in Definition 10, we shall now assume that the vectors $\pi_{kl}{}^{g_\nu}$ are also defined when F_k is empty, not uniquely but subject to the condition that their first components form a stochastic vector.

Suppose now that we are dealing with a P-net L which is normal relative to $[e, g_1, \ldots, g_h]$. Then, assuming that the numbers $a'_{kl}(g_\nu)$ are also defined when formula (8) becomes meaningless, we can form a matrix $\mathbf{A}_\nu = (a'_{kl}(g_\nu))$.

It is clear that $\mathbf{A}_\nu$ is strictly speaking not uniquely defined, but this ambiguity pertains only to an inessential aspect of the problem — nonattainable classes of states, and so it does not inconvenience us.

Using the matrices $\mathbf{A}_1, \mathbf{A}_2, \ldots, \mathbf{A}_h$, we now define new matrices $\mathbf{B}_1, \mathbf{B}_2, \ldots, \mathbf{B}_h$ and $\mathbf{B}'_1, \mathbf{B}'_2, \ldots, \mathbf{B}'_h$, all stochastic matrices of order k_1. The entries $b_{kl}(g_\nu)$ of $\mathbf{B}_\nu$ are defined by

$$b_{kl}(g_\nu) = \frac{a'_{kl}(g_\nu)}{\sum_{l=1}^{k_1} a'_{kl}(g_\nu)}, \tag{9}$$

and the entries $b'_{kl}(g_\nu)$ of $\mathbf{B}'_\nu$ by

$$b'_{kl}(g_\nu) = \frac{a'_{k_1+k,l}(g_\nu)}{\sum_{l=1}^{k_1} a'_{k_1+k,l}(g_\nu)}. \tag{10}$$

It is clear that our definitions of $\mathbf{B}_k$ and $\mathbf{B}'_k$ are legitimate only if the denominators in (9) and (10) do not vanish. We shall therefore assume that this is indeed the case whenever considering these matrices

We now define a stochastic vector $\widehat{\pi_e}$ as in the preceding section: (1) add together the entries in the first row of $\mathbf{L}_e$ whose second subscripts characterize a state of L corresponding to the same output word; (2) normalize the first k_1 components.

Definition 11. Let $\mathcal{A} = [\Sigma, S, \mathbf{M}_h{}^k, \pi_0, \alpha]$ be an FSA and L a P-net over $\mathcal{B}$ which is normal relative to $[e, g_1, \ldots, g_h]$. We shall say that L *realizes* $\mathcal{A}$ relative to $[e, g_1, \ldots, g_h]$ if

$$\begin{gathered} \pi_0 = \widehat{\pi_e}; \\ \forall \nu (\mathbf{M}_{\sigma_\nu} = \mathbf{B}_\nu = \mathbf{B}'_\nu), \quad 1 \leqslant \nu \leqslant h. \end{gathered} \tag{11}$$

A realization of $\mathcal{A}$ is said to be *synchronous* if the states defining an output word α with $\nu_\alpha = 0$ are nonattainable relative to $[e, g_1, \ldots, g_h]$, *asynchronous* otherwise.

Remark. Definitions 10 and 11 are not the only possible rigorous explications for our intuitive conceptions of a normal P-net L and a realization of $\mathcal{A}$. One can visualize a more general definition, which is nevertheless quite reasonable at least in an abstract context. This definition of the normality of L is based on specifying a system of input words

$[e, g_1, \ldots, g_h]$ and a system of finite sets of states of L, $[F_1, F_2, \ldots, F_m, F_{m+1}, \ldots \ldots, F_{2m}]$, satisfying the conditions:

(1) $\forall i \forall j((i \neq j) \supset (F_i \cap F_j \approx \varnothing))$;

(2) $L_{ds} \in (F_1 \cup F_2 \cup \ldots \cup F_m \cup F_{m+1} \cup \ldots \cup F_{2m})$, where L_{ds} is the class of states attainable relative to $[e, g_1, \ldots, g_h]$:

(3) all the components of the vector $\pi_{kl}{}^{g_v}$ are equal, $1 \leqslant k, l \leqslant 2m$;

(4) $((k \leqslant m) \& (k \neq s)) \supset (a'_{k,m+s}(g_v) = 0)$;

(5) $((k \leqslant m) \& (k \neq s)) \supset (a'_{m+k,m+s}(g_v) = 0)$.

Of course, this definition depends not only on the system of input words but also on a specified system of finite sets of states F_i; the latter system may be entirely unrelated to the outputs of L, as a result of which the notion of output looses its special significance. The resulting definition of a realization of $\mathcal{A}$ has an advantage over Definition 10: L may also be permitted as a realization of $\mathcal{A}$ when all the F_i are singletons. Apart from trivial cases, this is impossible with our original definition.

Theorem 1. A synthesis basis $\mathcal{B} = [\mathcal{A}_\neg, \mathcal{A}_\&, \mathcal{A}_\vee, \mathcal{A}_t, \mathcal{A}]$, where $\mathcal{A}$ is an element with exactly two internal states, is complete if and only if there is an input word y for $\mathcal{A}$ such that all entries of the matrix $\mathbf{M}_y$ are positive.

Proof. Let $\mathcal{B}$ be complete. Let $\mathcal{A}'$ be the FSA specified through alphabets $\Sigma = \{\sigma\}$, $S = \{s_1, s_2\}$, transition matrix

$$\mathbf{M}_\sigma = \begin{pmatrix} \frac{1}{2} & \frac{1}{2} \\ \frac{1}{2} & \frac{1}{2} \end{pmatrix},$$

initial distribution $\pi_0 = (1, 0)$ and final state vector $\alpha = \begin{pmatrix} 0 \\ 1 \end{pmatrix}$. By virtue of completeness, $\mathcal{B}$ admits a realization of an FSA arbitrarily close to $\mathcal{A}'$. In particular, we can construct a P-net L over $\mathcal{B}$ which realizes an FSA $\widehat{\mathcal{A}} = [\widehat{\Sigma}, \widehat{S}, M_1{}^2, \widehat{\pi}_0, \widehat{\alpha}]$ relative to $[e, g]$ with

$$\widehat{\Sigma} = \Sigma; \quad \widehat{S} = S; \quad \widehat{\pi}_0 = \pi_0; \quad \widehat{\alpha} = \alpha;$$

$$\widehat{\mathbf{M}}_\sigma = \begin{pmatrix} p & q \\ p' & q' \end{pmatrix},$$

where

$$\max\left(\left|\frac{1}{2} - p\right|, \left|\frac{1}{2} - p'\right|\right) < \frac{1}{16}.$$

We may thus condense the matrix $\mathbf{L}_g$ with respect to the classes F_1, F_2 of attainable states of L, the result being the matrix $\widehat{\mathbf{M}}_\sigma$.

Suppose further that L has $n + k_0$ vertices, k_0 of which are input vertices. This means that every state η_i of L is a word of length n:

$$\eta_i \cong s_{i_1} s_{i_2} \ldots s_{i_n},$$

where s_{i_j}, $j=1, 2, \ldots, n$, is an internal state of the basis element attached to the $j+k_0$-th vertex of the underlying graph of L. To simplify matters, we shall suppose that $\mathcal{A}$ is attached only to the vertices k_0+1, $k_0+2, \ldots, \ldots, k_0+k$. Now let η_i be a state attainable relative to $[e, g]$ which lies in F_1. Let $t_1, t_2, \ldots, t_{m_1}$ be the indices of all states of L belonging to F_1, and define $u_1, u_2, \ldots, u_{m_2}$ similarly for F_2. Then, by virtue of our assumptions,

$$\begin{cases} \frac{1}{2}-\frac{1}{16}<a_{it_1}(g)+a_{it_2}(g)+\ldots+a_{it_{m_1}}(g)<\frac{1}{2}+\frac{1}{16}; \\ \frac{1}{2}-\frac{1}{16}<a_{iu_1}(g)+a_{iu_2}(g)+\ldots+a_{iu_{m_2}}(g)<\frac{1}{2}+\frac{1}{16}. \end{cases} \tag{12}$$

Consequently, among the entries $a_{it_1}(g)$, $a_{it_2}(g), \ldots, a_{it'_{m_1}}(g)$ there must be one, $a_{ij}(g)$ say, such that

$$\frac{\frac{1}{2}-\frac{1}{16}}{m_1} \ngtr a_{ij}(g) \ngtr \frac{1}{2}+\frac{1}{16}. \tag{13}$$

This means that we can find an input word x' for the element $\mathcal{A}$ such that

$$l(x')=l(g);$$

$$\mathbf{M}_{x'}=\begin{pmatrix} a_{11}(x') & a_{12}(x') \\ a_{21}(x') & a_{22}(x') \end{pmatrix};$$

$$\left(\frac{\frac{1}{2}-\frac{1}{16}}{m_1} \ngtr a_{11}(x') \ngtr \sqrt[k]{\frac{1}{2}+\frac{1}{16}}\right) \vee \left(\frac{\frac{1}{2}-\frac{1}{16}}{m_1} \ngtr a_{21}(x') \ngtr \sqrt[k]{\frac{1}{2}+\frac{1}{16}}\right).$$

The word we want is clearly x' if both members of the disjunction are true. However, it follows at once from (12) only that one of them is true. We therefore assume the truth of

$$\frac{\frac{1}{2}-\frac{1}{16}}{m_1} \ngtr a_{11}(x') \ngtr \sqrt[k]{\frac{1}{2}+\frac{1}{16}}.$$

We now assume that one of the first k letters of the word η_i is s_1. * We shall see later that this assumption is necessary to make the proof general. Define a partial order relation $\lesseqqgtr$ on the states of L: let $\eta_u \equiv s_{u_1} s_{u_2} \ldots s_{u_n}$ and $\eta_v \equiv s_{v_1} s_{v_2} \ldots s_{v_n}$; then $\eta_u \lesseqqgtr \eta_v$ if and only if $\forall\, l(i_l \leqslant j_l)$, $l=1, 2, \ldots, k$.

If $\eta_u \lesseqqgtr \eta_v$ and $\neg(\eta_v \lesseqqgtr \eta_u)$, we shall also write $\eta_u < \eta_v$.

It is now easily seen that there exists an input word x'' for $\mathcal{A}$ such that the entry $a_{21}(x'')$ of $\mathbf{M}_{x''}$ is positive provided $\eta_j < \eta_i$ or η_i is not comparable with η_j. But then one of the matrices $\mathbf{M}_{x''x'}$, $\mathbf{M}_{x'x''x'}$ is positive (i. e., none of its entries vanish). We therefore consider the case that $\eta_i \lesseqqgtr \eta_j$. Since $a_{ij}(g)$ satisfies (13), there is a state η_i such that

* If the second member of the disjunction is assumed to be true, a similar assumption must be made for s_2.

$$\left(0 < a_{ij}(g) < \frac{1}{2} + \frac{1}{16}\right) \& (\eta_i < \eta_{i'}) \vee (\eta_{i'} < \eta_i) \vee \neg(\eta_i \lesseqgtr \eta_{i'}). \quad (14)$$

Hence we again conclude that there exists an input word y for which all the entries of $\mathbf{M}_y$ are positive, provided the following disjunction is true:

$$(\eta_{i'} < \eta_i) \vee \neg(\eta_i \lesseqgtr \eta_{i'}).$$

We therefore assume that $\eta_i < \eta_{i'}$.

Now, since η_i is attainable, it follows from (14) that the state $\eta_{i'}$ is also attainable. If we compare η_i and $\eta_{i'}$ as to the number of occurrences of s_2 in their prefixes of length k, we see that the number for $\eta_{i'}$ is larger. Since $\eta_{i'}$ is attainable, we may essentially repeat all the arguments given for η_i, finally determining either a word y meeting our requirements, or an attainable state $\eta_{i''}$ such that $\eta_{i'} < \eta_{i''}$.

It is clear that this process must terminate in a finite number m of steps, $m \leqslant k+1$, ultimately producing the required input word y. In fact, if no such word y is found up to and including the k-th step, a state whose first k letters are s_2 is attainable. Thus there exists a state η_w, such that

$$\neg(\eta_w \lesseqgtr \eta_{w'}) \& \left(\frac{\frac{1}{2} - \frac{1}{16}}{\max(m_1, m_2)} \ngtr a_{ww'}(g) \ngtr \frac{1}{2} + \frac{1}{16}\right).$$

Consequently, we can find a word x'' such that the entry $a_{22}(x'')$ of $\mathbf{M}_{x''}$ satisfies the inequalities $0 < a_{22}(x'') < 1$.

The second row of the matrix $\mathbf{M}_{x''}$ consists of positive entries, and so one of the matrices $\mathbf{M}_{x''x'}$ or $\mathbf{M}_{x'x''}$ is positive.

We have not considered the second member of the disjunction, i. e., the case that the relation

$$\frac{\frac{1}{2} - \frac{1}{16}}{m_1} \ngtr a_{21}(x') \ngtr \sqrt[k]{\frac{1}{2} + \frac{1}{16}}$$

is true. This is unnecessary, since the reasoning is exactly the same. This proves the first part of the theorem. We now proceed to prove the second part.

Let y be an input word such that the matrix $\mathbf{M}_y$ is positive. Then, using the constructively provable formula

$$\neg((a_1=0) \& (a_2=0) \& \ldots \& (a_n=0)) \supset ((a_1 \neq 0) \vee \vee (a_2 \neq 0) \vee \ldots \vee (a_n \neq 0)),$$

where $a_1, a_2, \ldots, a_n$ are CRNs, one easily proves that

$$\exists x((l(x) \leqslant 3) \& (a_{11}(x) a_{12}(x) a_{21}(x) a_{22}(x) > 0)).$$

For convenience, let us assume that

$$\mathbf{M}_x = \begin{pmatrix} a & b \\ c & d \end{pmatrix}$$

and

$$l(x) = 1.$$

Construct the P-net L shown in Figure 4. Its states are the words

$$\eta \leftrightharpoons s_{1+\alpha_1}\, s_{1+\alpha_2}\, s_{1+\alpha_3}\, s_{1+\alpha_4},$$

where $\alpha_i = 0$ or $\alpha_i = 1$, $i = 1, 2, 3, 4$ For example, if $\alpha_1 = 1$, $\alpha_2 = 0$, $\alpha_3 = 0$, $\alpha_4 = 1$, then L is in a state η such that the elements attached to its second and fifth vertices are in final states, the other elements in nonfinal states. Of course, we may consider sequences of values α_i instead of states of L. Arranging the states of L as indicated below Table 1, we get the transition matrix $\mathbf{L}_x$ of Table 1.

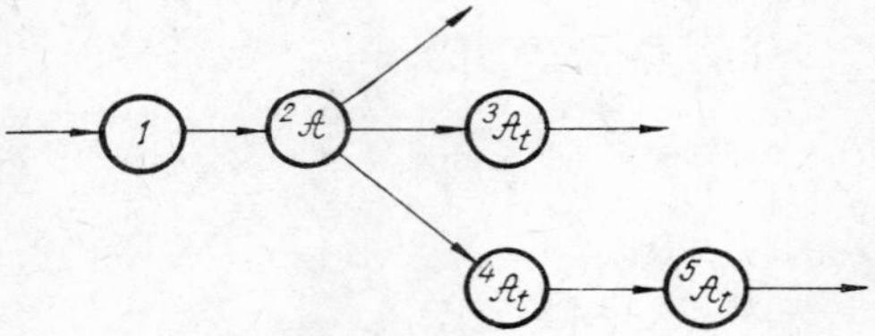

FIGURE 4.

It is easy to see that the matrix $(\mathbf{L}_x)^2$ may be condensed with respect to the class U_1 of all states from 1 through 4 and the class U_2 of all other states. The result is the matrix

$$\begin{pmatrix} bc & 1-bc \\ bc & 1-bc \end{pmatrix}.$$

It is now readily seen that the P-net L' of Figure 5* realizes the stochastic automaton $\mathcal{A}' = [\Sigma, S, \mathbf{M}_1^2, \pi_0, \alpha]$ relative to $[e, g]$, $e \leftrightharpoons g \leftrightharpoons xxxxxx$, where $\mathbf{M}'_\sigma = \begin{pmatrix} bc & 1-bc \\ bc & 1-bc \end{pmatrix}$, $\pi_0 = (bc,\ 1-bc)$.

If

$$\mathbf{M}_x = \begin{pmatrix} a & b \\ c & d \end{pmatrix},$$

the network emits the signals 0 and 1 with probabilities bc and $1-bc$, respectively. If $l(x) = 2$ ($l(x) = 3$), each of the vertices 3, 4, 5 in the network

* The right-hand section of the network realizes the boolean function $x_1\bar{x}_2x_3 \vee \bar{x}_1x_2\bar{x}_3$.

of Figure 4 may be replaced by a chain of two (three) delays. In that case $\mathcal{A}'$ is realized by the P-net obtained from that of Figure 5 by a similar transformation of the vertices 3, 4, 5. There is no difficulty in constructing appropriate pairs of input words $[e, g]$ for these cases. We only note here that when $l(x)=2$ $(l(x)=3)$ we have $l(e)=l(g)=8$ $(l(e)=l(g)=10)$.

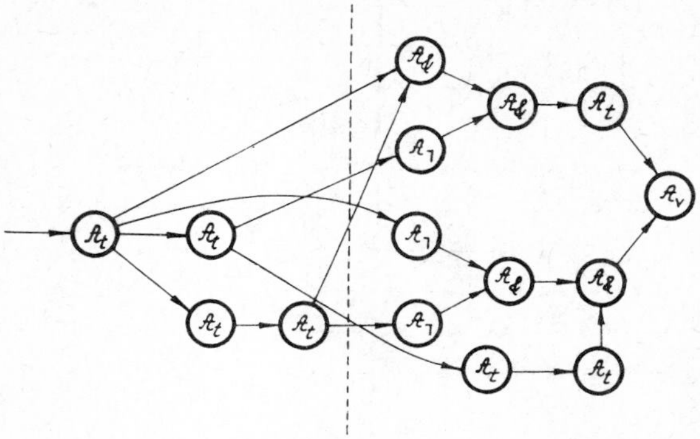

FIGURE 5.

The reader will probably have observed that this realization of $\mathcal{A}'$ is not standard, i. e., it does not agree with Definition 11. However, it possesses an important property for our purposes: it may be treated as a basis element emitting output 1 if and only if it is in a final state. One should not forget, though, that this realization of $\mathcal{A}'$ involves a "time dilatation" $\tau=ct$, where $c=l(g)$.

Now suppose we are given an FSA $\mathcal{A}$ which is to be realized with arbitrary tolerance $\varepsilon=2^{-\mu}$. Using the technique described in the proof of Proposition 4, we can construct a realization (with the prescribed tolerance) for each row of the matrix $\mathbf{M}_{\sigma_i}$, $i=1,2,\ldots,m$.

Using the elements $\mathcal{A}_t$, we can construct these realizations relative to the same pair of input words $[e, g]$, $e \supseteq g$.

Thus, we now have P-nets which realize the rows of the matrices $\mathbf{M}_{\sigma_i}$ relative to $[e, g]$. Let S_{ij} denote the P-net realizing the j-th row of $\mathbf{M}_{\sigma_i}$. If $\mathcal{A}$ has n internal states, we construct mn P-nets as shown in Figure 6.

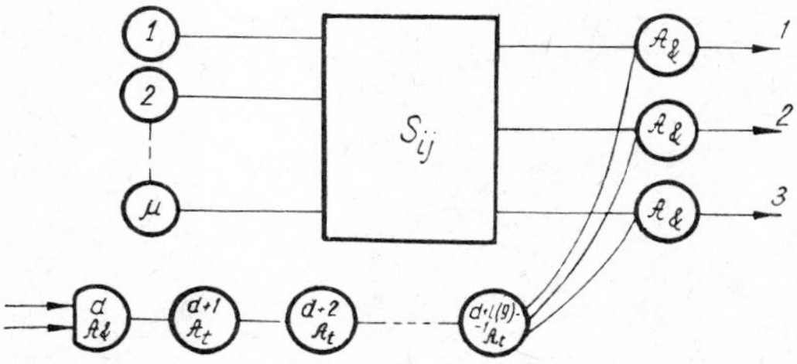

FIGURE 6.

TABLE 1.

	1	2	3	4	5	6	7	8	9	10	11	12	13	14	15	16
1					*a*						*b*					
2			*b*			*a*										
3		*c*													*d*	
4										*c*						*d*
5					*a*						*b*					
6					*a*						*b*					
7			*b*			*a*										
8			*b*			*a*										
9					*a*						*b*					
10			*b*			*a*										
11		*c*													*d*	
12										*c*						*d*
13		*c*													*d*	
14		*c*													*d*	
15										*c*						*d*
16										*c*						*d*

1) 0100; 2) 0110; 3) 1001; 4) 1011; 5) 0000; 6) 0001; 7) 0010; 8) 0011; 9) 0101; 10) 0111; 11) 1000; 12) 1010; 13) 1100; 14) 1101; 15) 1110; 16) 1111.

It is readily shown that there exists an input word f, $l(f)=l(e)=l(g)$, such that the P-net of Figure 6 realizes the j-th row of $\mathbf{M}_{\sigma_i}$ relative to $[f, f]$, provided of course that S_{ij} realizes this row relative to $[e, g]$, $e \mathrel{\bar{\bar{\circ}}} g$.

In order to avoid obscuring the exposition with technical details, we shall treat the case $m=2, n=3$. Then the P-net realizing $\mathscr{A}$ has the structure shown in Figure 7. The P-net S appearing as a component in Figure 7 realizes the initial distribution of $\mathscr{A}$ relative to $[e, g]$ and its structure is shown in Figure 8. Before defining the system of input words $[e', g_1, g_2]$ relative to which this P-net realizes $\mathscr{A}$, we need some new notation. We denote the word $\alpha_{t_1}\alpha_{t_2}\ldots\alpha_{t_\mu}$ 000 by ζ_t, the word $\alpha_{t_1}\alpha_{t_2}\ldots\alpha_{t_\mu}$ 100 by ζ'_t, the word $\alpha_{t_1}\alpha_{t_2}\ldots\alpha_{t_\mu}$ 010 by $\widehat{\zeta_t}$, and the word $\alpha_{t_1}\alpha_{t_2}\ldots\alpha_{t_\mu}$001 by $\widehat{\zeta'_t}$. Here $\alpha_{t_1}\alpha_{t_2}\ldots\alpha_{t_\mu}$ represents the t-th letter (counting from the beginning) of the word e. The input alphabet of the P-net in Figure 7 will be $Z=(\zeta_1, \ldots, \zeta_{l(e)}, \zeta'_1, \ldots, \zeta'_{l(e)}, \widehat{\zeta_1}, \ldots, \widehat{\zeta_{l(e)}}, \widehat{\zeta'_1}, \ldots, \widehat{\zeta'_{l(e)}})$. We now define $[e', g_1, g_2]$ by

$$e' \mathrel{\bar{\bar{\circ}}} \zeta_1 \ldots \zeta_{l(e)}\zeta'_1\zeta_2 \ldots \zeta_h;$$

$$g_1 \mathrel{\bar{\bar{\circ}}} \widehat{\zeta_1}\zeta_2 \ldots \zeta_{l(e)}\zeta_1 \ldots \zeta_h;$$

$$g_2 \mathrel{\bar{\bar{\circ}}} \widehat{\zeta'_1}\ \zeta_2 \ldots \zeta_{l(e)}\zeta_1 \ldots \zeta_h; \quad h \leqslant l(e).$$

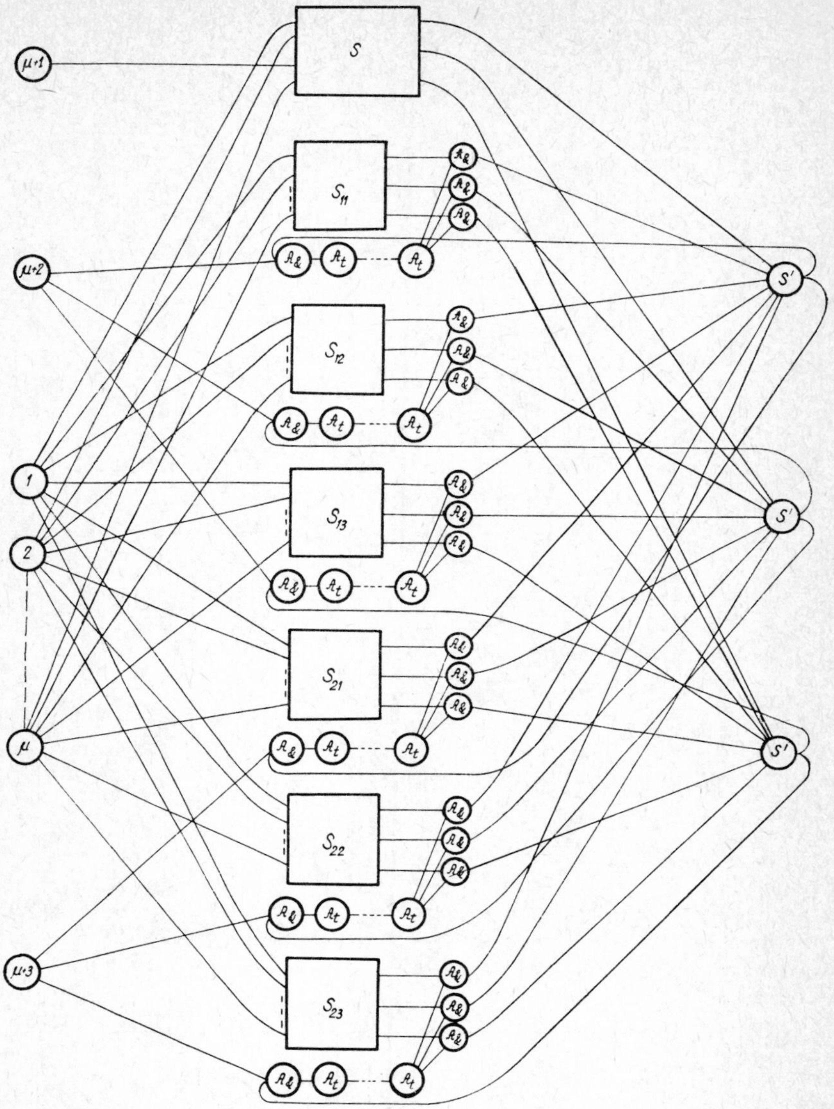

FIGURE 7.

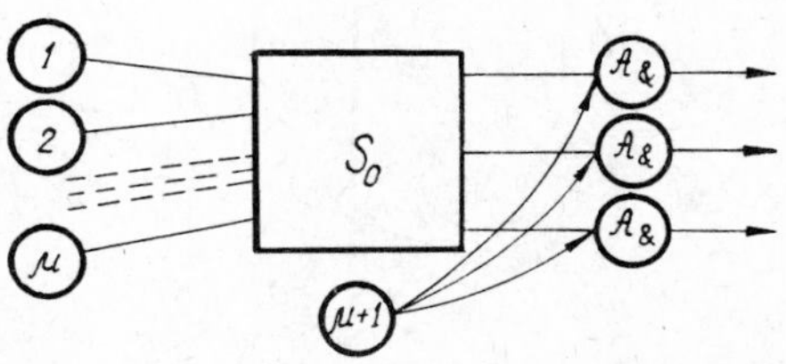

FIGURE 8.

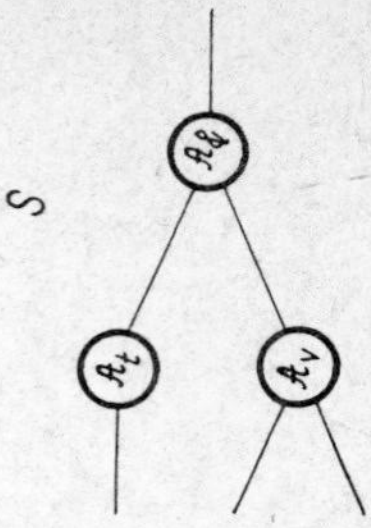

FIGURE 10.

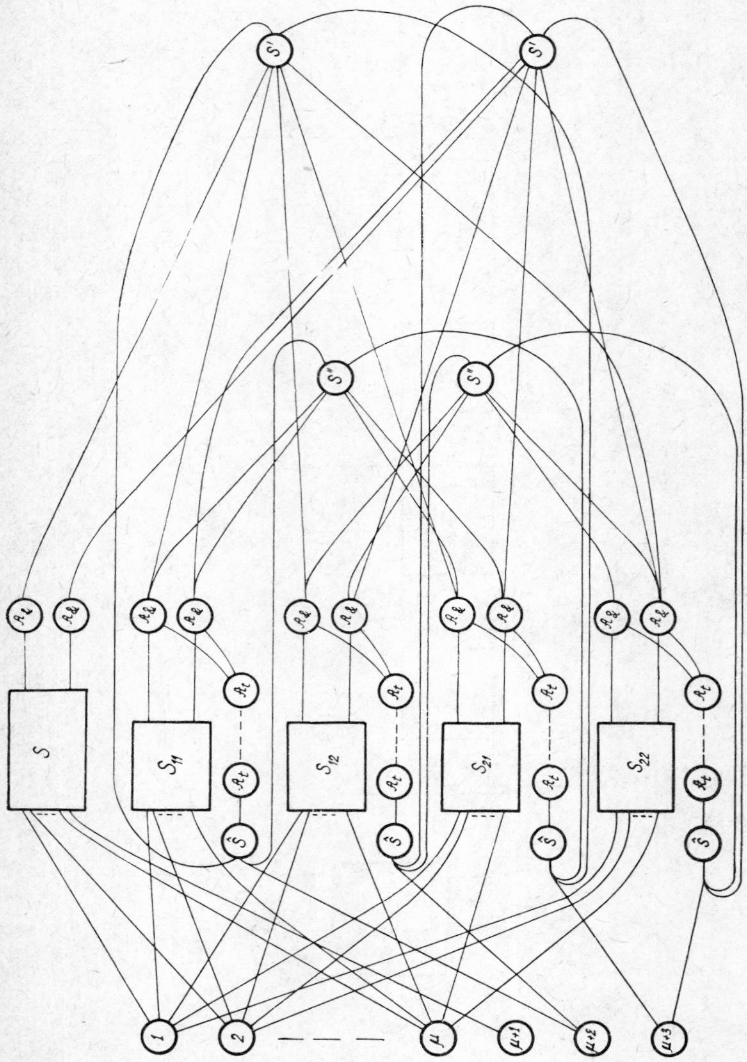

FIGURE 9.

Here h denotes the depth of the S'. The sets of states F_1, F_2, F_3 for our P-net, which represent the internal states of $\mathcal{A}$, are defined as follows. F_1 is the set of states corresponding to 1 at the output of S'_1, F_2 the set of states corresponding to output 1 of S'_2, and F_3 those corresponding to output 1 of S'_3. Note that the states corresponding to output 1 of all three P-nets S'_1, S'_2, S'_3 or of two of the three are nonattainable relative to $[e', g_1, g_2]$. Consequently, the classes of attainable states F_1, F_2, F_3 are pairwise disjoint. Also nonattainable relative to $[e', g_1, g_2]$ are all states corresponding to zero output of all three P-nets S'_1, S'_2, S'_3. Thus every attainable state of our P-net relative to $[e', g_1, g_2]$ belongs to the set $F_1 \cup F_2 \cup F_3$. Thus the P-net of Figure 7 indeed realizes the automaton $\mathcal{A}$ with tolerance ε.

The proof of the following theorem is analogous.

Theorem 2. The synthesis basis $\mathcal{B} = [\mathcal{A}_{\neg}, \mathcal{A}_{\&}, \mathcal{A}_{\vee}, \mathcal{A}_t, \mathcal{A}]$ *of Theorem 1 is strongly complete if and only if there is a word* y *such that the matrix* $\mathbf{M}_y$ *is positive.*

The proof is omitted; in order to help the reader reconstruct the proof, we illustrate the general structure of the realization of an FSA with two inputs and two internal states in Figures 9 and 10. It is evident that for practical implementation of the P-net one must provide a warning line for the eventuality that the network has performed a false transition and the preceding input word is to be repeated. We have disregarded this situation in our theoretical treatment of asynchronous realizations.

NOTES

The beginnings of the theory of structural synthesis of FSA may be traced back to many authors, of whom we mention Skhirtladze (1961), Davis (1961), Gill (1962), Tsertsvadze (1963, 1964), Chentsov (1967) and Knast (1967). The contributions of Davis, Gill, Chentsov and Knast share a remarkably general approach to the theory, all of them basing the structural synthesis of autonomous FSA on the decomposition of stochastic matrices. If such a decomposition is available, one can reduce the synthesis of an autonomous (and, in the final analysis, nonautonomous) FSA to the synthesis of an FDA with random input.

The conception of structural synthesis set forth here is closely bound up with the work of Makarevich (1968, 1969a, b, 1971). As a matter of fact, Makarevich was the first to formulate certain principles that any general theory of structural synthesis of FSA must obey; prior to this we find only a few isolated methods for structural realization of certain types of FSA. Makarevich was also the first to pose the problem of the completeness of bases for structural synthesis. He adheres to the principles of classical mathematics, so that our approach is necessarily different. But this is neither the only difference nor the most important one. In fact, Makarevich's conception differs from ours in the very interpretation of structural realization. He devotes almost all his attention to asynchronous realization of stochastic vectors and is not at all interested in synchronous realization.

It should be mentioned that the idea of asynchronous realization is entirely due to Makarevich and Giorgadze. We use it in a somewhat modified form, departing from Makarevich's definitions.

Underlying the most important definitions in this chapter is the binary principle for the functioning of basis elements. Of course, it is not difficult to do away with this principle. The question is only how interesting our definitions seem to the receptive reader. Today comparatively little hardware operates with ternary signals. Since we are not considering questions of structural synthesis as divorced from reality, this aspect of the problem should be pondered before even attempting to develop a more general conception of structural synthesis.

Podnieks has carried out a comparative analysis of the two definitions of completeness. He has found bases which are strongly but not weakly complete. In addition, he has shown that a complete basis $\mathcal{B}$ will also be strongly complete if it admits realization of the FDA $\mathcal{A} = [\Sigma, S, \mathbf{M}_4^3, \alpha]$ with

$$\mathbf{M}_{\sigma_1} = \begin{pmatrix} 1 & 0 & 0 \\ 0 & 1 & 0 \\ 0 & 1 & 0 \end{pmatrix}; \quad \mathbf{M}_{\sigma_2} = \mathbf{M}_{\sigma_4} = \begin{pmatrix} 0 & 1 & 0 \\ 1 & 0 & 0 \\ 1 & 0 & 0 \end{pmatrix};$$

$$\mathbf{M}_{\sigma_3} = \begin{pmatrix} 1 & 0 & 0 \\ 0 & 0 & 1 \\ 0 & 0 & 1 \end{pmatrix}; \quad \alpha = \begin{pmatrix} 0 \\ 0 \\ 1 \end{pmatrix}.$$

It is noteworthy that our definitions of completeness involve no conditions regarding the possible realization of any one FSA.

An alternative approach to the concept of realization of a stochastic vector and completeness of the basis, using definitions other than Definitions 8 and 9, would be interesting. The most natural definitions are probably the following.

Definition 8'. We shall say that a P-net L *ε-realizes* π_0 relative to $[e, g]$ if L is normal relative to $[e, g]$ and

$$\begin{cases} \omega(\pi_0 - \widehat{\pi_e}) \not> \varepsilon; \\ \omega(\pi_0 - \widehat{\pi_g}) \not> \frac{1}{2}\varepsilon; \\ \varkappa(\widehat{L_g}) \not> \frac{1}{2}\varepsilon. \end{cases}$$

Definition 9'. A basis $\mathcal{B}$ is said to be *complete* if every stochastic vector is ε-realizable in it for every ε.

Chapter IX

ACCURACY AND STABILITY OF STRUCTURAL REALIZATION

In this chapter we investigate some problems concerning the accuracy with which the true values of the parameters of the basis elements are specified or known, and some questions pertaining to the stability of the parameters, i. e., their constancy in time. At the basis of an analysis of this kind lie certain practical considerations which may be ignored only by the theoretician. Essentially, the argument is as follows. In practice one is never provided with accurate values of the transition probabilities of the basis elements. The main reason for this is that these parameters are in principle not accurately measurable, even in situations where their values may objectively be specified with precision. Moreover, one can never be sure that the parameter in question does indeed have a fixed value, and does not vary over a period of time. In the final analysis, therefore, no actual realization of a stochastic vector (or FSA) can justifiably claim to be absolutely accurate. This reservation by no means implies that any construction based on an initial idealization, involving the assumption that the basis elements may be realized with absolutely stable parameters and accurately known values of the parameters, is devoid of any real meaning. This is by no means the case. The truth is that this idealization is merely a first step toward the desired goal, and not an end in itself.

§1. RELATED P-NETS AND ACCURACY OF REALIZATION OF STOCHASTIC VECTORS

In this and the next section we shall be concerned with synthesis bases only of the following type:

$$\mathcal{B}=(\mathcal{A}_{\neg}, \mathcal{A}_{\&}, \mathcal{A}_{\vee}, \mathcal{A}_{t}, \mathcal{A}_1, \mathcal{A}_2, \ldots, \mathcal{A}_m),$$

where $\mathcal{A}_i$, $i=1, 2, \ldots, m$, are FSA with two inputs, two internal states and one positive transition matrix. To simplify the exposition, we shall assume that the positive matrix is that defined by σ_2, i. e., $(\mathbf{M}_i)_{\sigma_2}$, that the first state of $\mathcal{A}_i$ is the initial state and the second state the final state. The elements $\mathcal{A}_i$ may be equal.

Definition 1. Let $L=[G, \mathfrak{H}]$ be a P-net over a basis $\mathcal{B}$ and $L'=[G', \mathfrak{H}']$ a P-net over a basis $\mathcal{B}'$. We shall say that L' is related to L with tolerance ε if the following conditions are satisfied:

(1) $G\equiv G'$;

(2) $\mathcal{B}$ and $\mathcal{B}'$ contain the same number of elements;

(3) every vertex e of the graph G satisfies the formula

$$((\mathfrak{H}(e)=\mathcal{A}_i)\supset(\mathfrak{H}'(e)=\mathcal{A}'_i))\,\&$$
$$\&((\mathfrak{H}(e)=\mathcal{A}_\neg)\vee(\mathfrak{H}(e)=\mathcal{A}_\&)\vee(\mathfrak{H}(e)=\mathcal{A}_\vee)\vee$$
$$\vee(\mathfrak{H}(e)=\mathcal{A}_t)\supset(\mathfrak{H}(e)=\mathfrak{H}'(e));$$

(4) $\varepsilon=1-\max\,(\omega((\mathbf{M}_1)_{\sigma_2}-(\mathbf{M}'_1)_{\sigma_2}),\ \ldots,\ \omega((\mathbf{M}_m)_{\sigma_2}-(M'_m)_{\sigma_2}))$.

It is easy to see that the tolerance ε can never vanish, and its maximum value is 1.

Theorem 1. Given a basis $\mathcal{B}=(\mathcal{A}_\neg,\ \mathcal{A}_\&,\ \mathcal{A}_\vee,\ \mathcal{A}_t,\ \mathcal{A}_1)$, *an arbitrary pair of natural numbers* k, k', *and a stochastic vector* $\pi_0=(p_1, p_2 \ldots, p_l)$, *one can construct a* P-*net* L *with the following properties:*

(1) Relative to some system of words $[e, g]$, L *realizes a stochastic vector* $\pi'_0=(p'_1,\ p'_2,\ \ldots,\ p'_l)$ *such that* $\omega(\pi_0-\pi'_0)<2^{-k}$.

(2) Any P*-net* L' *related to* L *realizes relative to* $[e, g]$ *a stochastic vector* π''_0 *such that*

$$\omega(\pi_0-\pi''_0)<2^{-k},$$

provided the element $\mathcal{A}'_1$ *of the basis* $\mathcal{B}'$ *over which* L' *is synthesized satisfies the condition*

$$\omega((\mathbf{M}'_1)_{\sigma_2})<1-2^{-k'}. \tag{1}$$

Proof. We begin with a lemma

Lemma 1. Let Ω^r *be the field of* r*-independent Bernoulli trials associated with a probability field* $\Omega=(E, \mathfrak{B})$. *For any sequence of natural numbers* n, c_1, c_2, $\ldots$, c_l *such that*

$$\sum_{i=1}^{l} c_i=n\leqslant r,$$

there is in Ω^r *a system of mutually exclusive events* $A_0, A_1, \ldots, A_l$ *satisfying the conditions:*

(a) $\sum_{i=0}^{l} p(A_i)=1$;

(b) $p(A_0)<(n-1)(r+1)\varrho^r$, *where* $\varrho=\max\,(\mathfrak{B}(e_1),\ \mathfrak{B}(e_2))$;

(c) $p(A_i|\overline{A}_0)=\dfrac{c_i}{n}$, $i=1, 2, \ldots, l$.

Proof. Let B_m denote the event consisting of exactly those elementary events of Ω^r in which there are just m occurrences of $\xi_d e_1 \xi_d$. The event B_m clearly contains C_r^m elementary events. Each event B_m, $0<m<r$, may be split into $n+1$ mutually exclusive events $B_{m0}, B_{m1}, \ldots, B_{mn}$ such that

$$\begin{cases} p(B_{m1}) = p(B_{m2}) = \ldots = p(B_{mn}); \\ \nu(B_{m0}) = \operatorname{res}\left(\frac{C_r^m}{n}\right). \end{cases}$$

Define events A'_i, $i=1, \ldots, n$, by

$$A'_i \approx B_{1i} \cup B_{2i} \cup \ldots \cup B_{r-1,i}.$$

Clearly, the events A'_i are mutually exclusive and have the same probability. We now define the required system of events by

$$\begin{aligned} &A_0 \approx B_0 \cup B_{10} \cup \ldots \cup B_{r-1,0} \cup B_r; \\ &A_1 \approx A'_1 \cup A'_2 \cup \ldots \cup A'_{c_1}; \\ &A_2 \approx A'_{c_1+1} \cup \ldots \cup A'_{c_1+c_2}; \\ &\cdot \quad \cdot \quad \cdot \quad \cdot \quad \cdot \quad \cdot \quad \cdot \quad \cdot \quad \cdot \quad \cdot \\ &A_l \approx A'_{n-c_l+1} \cup \ldots \cup A'_n. \end{aligned}$$

It is easy to see that the events A_i are mutually exclusive, and the sum of their probabilities is unity. Since the events $B_0, B_{10}, \ldots, B_{r-1,0}, B_r$ are mutually exclusive, we have

$$p(A_0) = p(B_0) + p(B_{10}) + \ldots + p(B_{r-1,0}) + p(B_r).$$

But since

$$(\mathfrak{B}(e_1))^m \cdot (1 - \mathfrak{B}(e_1))^{r-m} \not> \varrho^r$$

for any m, $m=0, 1, \ldots, r$, we obtain the following estimate for $p(A_0)$:

$$p(A_0) \not> (n-1)\varrho^r(a_0 + a_1 + \ldots + a_r),$$

where

$$a_m = \frac{(\mathfrak{B}(e_1))^m (1 - \mathfrak{B}(e_1))^{r-m}}{\varrho^r},$$

$m=0, 1, \ldots, r$. Since the numbers a_m do not exceed 1, it follows from the above inequality that $p(A_0) \not> (n-1)(r+1)\varrho^r$. The truth of (c) follows from the easily proved equalities

$$\begin{aligned} p(A_i) &= c_i p(A'_1), \quad i = 1, 2, \ldots, l; \\ p(\overline{A_0}) &= np(A'_1). \end{aligned}$$

This completes the proof of our lemma.

Proof of Theorem 1. We may assume for simplicity's sake that the matrix $(\mathbf{M}_1)\sigma_2$ has the form

$$\begin{pmatrix} a & b \\ c & d \end{pmatrix}.$$

Then the P-net L_1 shown in Figure 11 realizes relative to the system of input words $[\sigma_2\sigma_2\sigma_1\sigma_1\sigma_1\sigma_1, \sigma_2\sigma_2\sigma_1\sigma_1\sigma_1\sigma_1]$ an FSA $\mathcal{A}$ with one input state σ and transition matrix

$$\mathbf{M}_\sigma = \begin{pmatrix} bc & 1-bc \\ bc & 1-bc \end{pmatrix}.$$

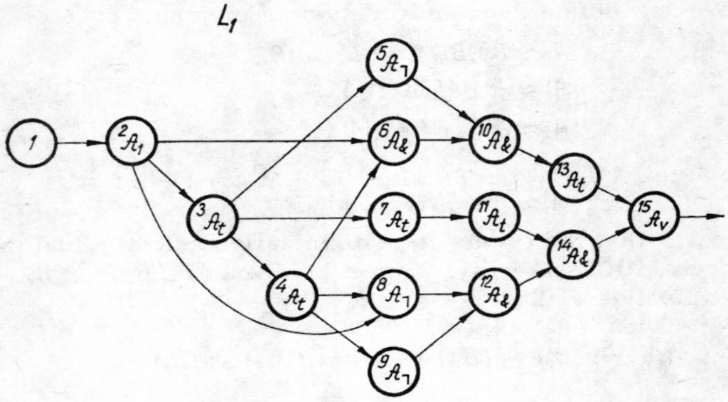

FIGURE 11.

Now let $c_1, \ldots, c_l, n$ be a system of natural numbers such that

$$\sum_{i=1}^{l} c_i = n; \tag{2}$$

$$\forall i\left(\left|p_i - \frac{c_i}{n}\right| < 2^{-k-1}\right). \tag{3}$$

Let r' be an integer such that $r' \geqslant \left(\frac{4}{\ln^2 \varrho'} + k + 1 + \ln n\right)^2$, where $\varrho' = \max\ (bc,\ 1-bc,\ 1-4^{-k'})$, and construct a P-net L_2 as shown in Figure 12.

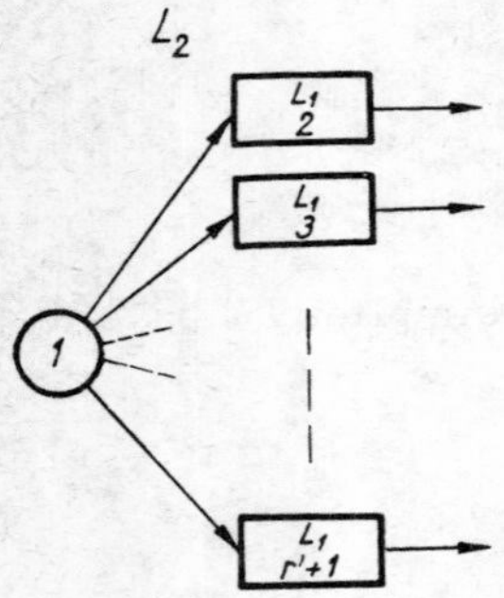

FIGURE 12.

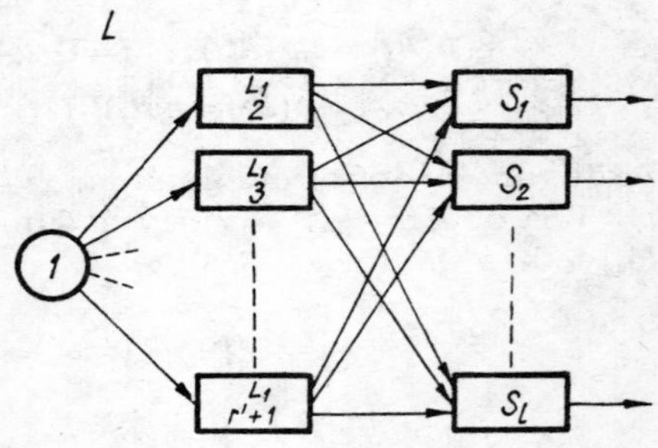

FIGURE 13.

Obviously, the output sequences $(\alpha_1, \alpha_2, \ldots, \alpha_r)$ of L_2, realized by the system of input words $[\sigma_2\sigma_2\sigma_1\sigma_1\sigma_1\sigma_1, \sigma_2\sigma_2\sigma_1\sigma_1\sigma_1\sigma_1]$, may be interpreted as elementary events in the field Ω^r of r-independent Bernoulli trials associated with a probability field $\Omega=(E, \mathfrak{B})$, where $\mathfrak{B}(e_1)=bc$. By Lemma 1, the output sequences of L_2 may be partitioned into classes $A_0, A_1, \ldots, A_l$ satisfying conditions (a)—(c) (with r replaced by r'). From the elementary events in A_0 we form a system of mutually exclusive events $A_{01}, A_{02}, \ldots, A_{0l}$ such that

(i) $\mu(A_{01})=\mu(A_{02})=\ldots=\mu(A_{0l-1})$;

(ii) $\mu(A_{0l})<2l$.

It is easy to see that for any i, $i=1, \ldots, l-1$,

$$p(A_{0i}) \not> E\left(\frac{(n-1)(r+1)}{l}\right)\varrho'^2.$$

We now define a new system of pairwise disjoint classes of sequences:

$$C_i \approx A_i \cup A_{0i}, \quad i=1, \ldots, l.$$

The transition probability from a state η of the B-net L_2 to another state η', $\eta' \in C_i$, when the input to L_2 is the word $\sigma_2\sigma_2\sigma_1\sigma_1\sigma_1\sigma_1$, is equal to $p(C_i)$. By Bayes' theorem, together with conditions (i), (ii),

$$p(C_i)=p(\overline{A}_0)p(A_i|\overline{A}_0)+p(A_0)p(A_{0i}|A_0)=$$
$$=(1-p(A_0))\frac{c_i}{n}+p(A_0)d_i,$$

where $d_i=p(A_{0i}|A_0) \not> 1$.

Consequently, we obtain the following bounds for $p(C_i)$:

$$\frac{c_i}{n}-2^{-k-1}\cdot\frac{c_i}{n} \not> p(C_i) \not> \frac{c_i}{n}+2^{-k-1}\left(1-\frac{c_i}{n}\right).$$

Hence it follows that

$$\forall i(|p_i-p(C_i)|<2^{-k}).$$

Let S_i, $i=1, 2, \ldots, l$, denote a P-net realizing the boolean function $f_i(x_1, x_2, \ldots, x_{r'})$ with truth domain C_i, and construct a P-net L as in Figure 13. We see that L realizes (relative to the same system of input words) a stochastic vector π'_0 satisfying condition (1) in the statement of the theorem. We claim that it also satisfies condition (2), so that the proof will be complete. Indeed, if

$$(\mathbf{M}'_1)_{\sigma_2}=\begin{pmatrix} a' & b' \\ c' & d' \end{pmatrix},$$

then it follows from formula (1) that

$$\min(b'c', 1-b'c')>2^{-2k'}.$$

Thus, in the case of the related P-net L' we again have the field of r'-independent Bernoulli trials associated with a probability field $\Omega = (E, \mathfrak{B})$ such that

$$\max(\mathfrak{B}(e_1), \mathfrak{B}(e_2)) \not> \varrho'.$$

But then all our estimates for the components of the stochastic vector π'_0 remain in force for the components of the vector π''_0 realized by L', and the proof is complete.

Using Lemma 1, we can also prove the following

Theorem 1'. Given a basis $\mathcal{B} = (\mathcal{A}_\neg, \mathcal{A}_\&, \mathcal{A}_\vee, \mathcal{A}_t, \mathcal{A}_1)$, an arbitrary pair of natural numbers k, k', and a stochastic vector $\pi_0 = (p_1, p_2, \ldots, p_l)$, one can construct a P-net L with the following properties:

(1) The element $\mathcal{A}_1$ is attached to only one vertex of L.

(2) L realizes relative to some system of words [e, g] a stochastic vector $\pi'_0 = (p'_1, p'_2, \ldots, p'_l)$ such that $\omega(\pi_0 - \pi'_0) < 2^{-k}$.

(3) Every P-net L' related to L realizes relative to [e, g] a stochastic vector π''_0 such that $\omega(\pi_0 - \pi''_0) < 2^{-k}$, provided the element $\mathcal{A}'_1$ in the basis $\mathcal{B}'$ over which L' is synthesized satisfies the condition

$$\omega((\mathbf{M}'_1)_{\sigma_2}) < 1 - 2^{-k'}.$$

We shall not give the details of the proof, only sketching the construction of the required P-net.

We first construct a P-net L_1 as in Figure 11, and then a P-net L'_2 as shown in Figure 14, where L_{1i}, $i = 1, 2, 3, 4$, is a chain of $l(g) = 6$ delays, $r = 4$. L'_2 plays the same role here as L_2 in the proof of Theorem 1. The final step of the construction is essentially the same as in the proof of Theorem 1.

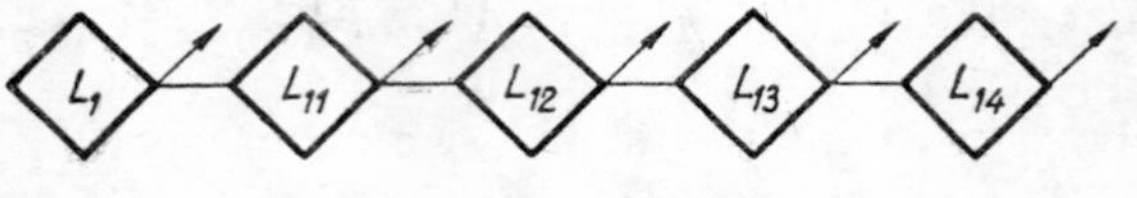

FIGURE 14.

If the realization is allowed to be asynchronous, Theorems 1 and 1' may be strengthened. We present the corresponding analog of Theorem 1 without proof.

Theorem 2. Given a basis $\mathcal{B} = (\mathcal{A}_\neg, \mathcal{A}_\&, \mathcal{A}_\vee, \mathcal{A}_t, \mathcal{A}_1)$, an arbitrary pair of natural numbers k, k', and a stochastic vector $\pi_0 = (p_1, p_2, \ldots, p_l)$ with rational components, one can construct a P-net L with the following properties:

(1) L realizes π_0 relative to some system [e, g].

(2) Every P-net L' related to L realizes π_0 relative to [e, g] with asynchronous degree less than 2^{-k}, provided the element $\mathcal{A}'_1$ of the basis $\mathcal{B}'$ underlying the synthesis of L' satisfies the condition $\omega((\mathbf{M}'_1)_{\sigma_2}) < 1 - 2^{-k'}$.*

* The asynchronous degree is defined as the maximum probability of a false transition, i.e., a transition to a state in F_{l+1}.

The net L of Theorem 2 is illustrated schematically in Figure 15, for the case $r'+1=4$, $l=3$. Each S_i, $i=1, 2, 3$, realizes a boolean function $f_i(x_1, x_2, x_3, x_4)$ such that

$$\begin{cases} f_1 \vee f_2 \vee f_3 \neq 1; \\ (i \neq j) \supset (f_i f_j = 0). \end{cases}$$

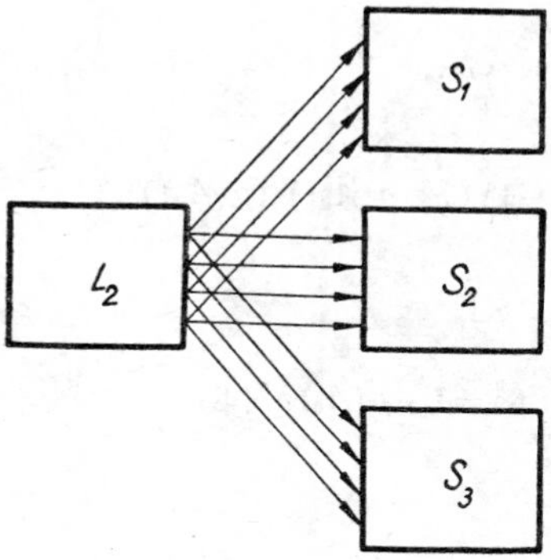

FIGURE 15.

Considering the net L, one can ask how many times the word g must be applied (on the average) to the input in order to take L from some state $\eta \in F_1 \cup \ldots \cup F_l$ to the class of states $F_1 \cup \ldots \cup F_l$, where l is the number of components of π_0. It is clear that the probability of a false transition is the same for all states η, namely $p(A_0)$, if g is applied to the input of L only once. Consequently, the required quantity is the mean of a certain random variable $\mathfrak{P}$ over a probability field $\Omega=(E, \mathfrak{B})$ with enumerable set E. The probability measure $\mathfrak{B}$ and random variable $\mathfrak{P}$ are defined by

$$\mathfrak{B}(e_i) = (1-p(A_0)) \cdot (p(A_0))^{i-1};$$
$$\mathfrak{P}(e_i) = i, \quad i=1, 2, \ldots$$

Thus, the mean and variance of $\mathfrak{P}$ (provided they exist) are given by

$$M(\mathfrak{P}) = \sum_{i=1}^{\infty} i(1-p(A_0))(p(A_0))^{i-1};$$

$$D(\mathfrak{P}) = \sum_{i=1}^{\infty} i^2(1-p(A_0))(p(A_0))^{i-1} - [M(\mathfrak{P})]^2.$$

We shall show that $M(\mathfrak{P})$ and $D(\mathfrak{P})$ exist and are equal to $\frac{1}{1-p(A_0)}$ and $\frac{p(A_0)}{(1-p(A_0))^2}$, respectively. Using elementary techniques of constructive mathematics, one can show that the sequence $\{\sqrt[n-1]{n}\}$ converges constructively to unity. Consequently, the series

$$\sum_{i=1}^{\infty} i(1-p(A_0))(p(A_0))^{i-1}$$

is absolutely convergent. This series may be expressed as the sum of two convergent series:

$$\sum_{i=1}^{\infty} i(1-p(A_0))(p(A_0))^{i-1} = \sum_{i=1}^{\infty} (1-p(A_0))(p(A_0))^{i-1} + \\ + \sum_{i=1}^{\infty} (i-1)(1-p(A_0))(p(A_0))^{i-1}.$$

Thus, we obtain the formula

$$M(\mathfrak{P}) = 1 + p(A_0)M(\mathfrak{P}),$$

whence it follows that

$$M(\mathfrak{P}) = \frac{1}{1-p(A_0)}. \tag{4}$$

The convergence of the series

$$\sum_{i=1}^{\infty} i^2(1-p(A_0))(p(A_0))^{i-1}$$

follows again from that of the sequence $\{\sqrt[n-1]{n}\}$. Express this series as the sum of two convergent series:

$$\sum_{i=1}^{\infty} i^2(1-p(A_0))(p(A_0))^{i-1} = \sum_{i=1}^{\infty} (2i-1)(1-p(A_0))(p(A_0))^{i-1} + \\ + \sum_{i=1}^{\infty} (i-1)^2(1-p(A_0))(p(A_0))^{i-1}.$$

Denoting the sum of the series

$$\sum_{i=1}^{\infty} i^2(1-p(A_0))(p(A_0))^{i-1}$$

by $M(\mathfrak{P}^2)$, we obtain from (4)

$$M(\mathfrak{P}^2) = 2M(\mathfrak{P}) - 1 + p(A_0)M(\mathfrak{P}^2).$$

Consequently,

$$M(\mathfrak{P}^2)=\frac{1+p(A_0)}{(1-p(A_0))^2}. \qquad (5)$$

Finally, using (4) and (5), we get

$$D(\mathfrak{P})=M(\mathfrak{P}^2)-(M(\mathfrak{P}))^2=\frac{p(A_0)}{(1-p(A_0))^2}.$$

Thus $D(\mathfrak{P})$ has the stipulated value.

We see that $1-M(\mathfrak{P})$ and $D(\mathfrak{P})$ may be made arbitrarily small. Thus the net L of Figure 15 functions "almost synchronously."

§2. DISTRIBUTION GENERATORS AND THE TIME-OPTIMAL PROBLEM

Let $\Omega_1, \Omega_2, \ldots, \Omega_c$ be a system of PFs, where the space of elementary events of each $\Omega_i=(E_i, \mathfrak{B}_i)$ consists of only two elements. A probability field $\Omega=(E, \mathfrak{B})$ will be called the product of the PFs Ω_i, denoted by $\Omega_1\times\Omega_2\times\ldots\times\Omega_c$, provided the following conditions hold:

(1) E is the set of all words

$$\delta e_{1i_1}\delta e_{2i_2}\ldots e_{ci_c}\delta,$$

where δ is a letter not occurring in any elementary event of a PF Ω_i.

(2) $$\mathfrak{B}(\delta e_{1i_1}\delta e_{2i_2}\ldots e_{ci_c}\delta)=\mathfrak{B}_1(e_{1i_1})\mathfrak{B}_2(e_{2i_2})\ldots\mathfrak{B}_c(e_{ci_c}).$$

Henceforth, unless otherwise stated, we shall deal only with products of probability fields Ω_i satisfying the condition

$$0<\mathfrak{B}_j(e_{ji_j})<1, \quad j=1, 2, \ldots, c.$$

Lemma 2. Let $\Omega=\Omega_1\times\Omega_2\times\ldots\times\Omega_c=(E, \mathfrak{B})$ be a PF such that $c=rd$ and for all natural numbers r', $r'<r$,

$$\Omega_{r'd+1}=\Omega_{r'd+2}=\ldots=\Omega_{r'd+d}.$$

Then, for any prescribed system of natural numbers $c_1, c_2, \ldots, c_l$, n such that

$$d^{h-1}<\sum_{i=1}^{l}c_i=n\leqslant d^h \quad (h\leqslant r), \qquad (6)$$

there is a system of mutually exclusive events $A_0, A_1, \ldots, A_l$ in Ω such that

(a) $\sum_{i=0}^{l} p(A_i) = 1$;

(b) $p(A_i | \overline{A}_0) = \frac{c_i}{n}, \quad i = 1, 2, \ldots, l$;

(c) $p(A_0) \not> (n-1)(\varrho^d(d+1))^r + (2\varrho^d)^r (2^{d-1}-1)^{h-1} r^{h-1}$, *where* $\varrho = \max_j (\mathfrak{B}_j(e_{j1}), 1 - \mathfrak{B}_j(e_{j1}))$.

Proof. For each i, $0 \leqslant i \leqslant r-1$, we define a random variable $\mathfrak{P}_i$ over Ω as follows: $\mathfrak{P}_i(e) = m$ if and only if $e \overline{\subseteq} \delta e_{1j_1} \delta \ldots \delta e_{drj_{dr}} \delta$ and exactly m words in the sequence $e_{id+1, j_{id+1}}, \ldots, e_{id+d, j_{id+d}}$ are graphically equal to the word $e_{id+1,1}$.

It is clear that the values of $\mathfrak{P}_i$ lie between 0 and d. Let A_{im} denote the event $\{\mathfrak{P}_i = m\}$. We define a new system of events as follows.

Let $\overline{q}$ be a vector with natural components $q_{i_0}, q_{i_1}, \ldots, q_{i_{r-1}}$ such that $\forall\, i (0 \leqslant q_i \leqslant d)$.

Let $B_{\overline{q}}$ denote the event

$$A_{0q_0} A_{1q_1} \ldots A_{r-1q_{r-1}}.$$

It is easily seen that the events $B_{\overline{q}}$ and $B_{\overline{q}'}$ are mutually exclusive if $\overline{q} \neq \overline{q}'$. We now confine attention to the events $B_{\overline{q}}$ for $\overline{q}$ having at least h components between 1 and $d-1$, i. e., such that

$$0 < q_i < d. \qquad (7)$$

Let us call a component q_i significant if it satisfies (7). Renumber the events of this class in some way and let C_j denote the j-th term in the resulting sequence of events, $j = 1, 2, \ldots, \nu$.

The events C_j possess an important property: if e and e' are elementary events contained in C_j, then $\mathfrak{B}(e) = \mathfrak{B}(e')$.

Since the number of elementary events contained in $B_{\overline{q}}$ is

$$\mu(B_{\overline{q}}) = C_d^{\,q_0} C_d^{\,q_1} \ldots C_d^{\,q_{r-1}},$$

it follows that $\forall\, j (\mu(C_j) \geqslant d^h)$.

Consequently, by the assumptions of the lemma, $\mu(C_j) \geqslant n$.

This means that C_j may be partitioned into $n+1$ mutually exclusive events $C_{j0}, C_{j1}, \ldots, C_{jn}$ such that:

(a) $\mu(C_{j0}) = \mathrm{res}\left(\frac{\mu(C_j)}{n}\right)$;

(b) $\mu(C_{j1}) = \mu(C_{j2}) = \ldots = \mu(C_{jn}) > 0$;

(c) $C_{j0} \cup C_{j1} \cup \ldots \cup C_{jn} \approx C_j$.

We now define a system of events $A'_0, A'_1, \ldots, A'_n$ by

$$A'_i \approx C_{1i} \cup C_{2i} \cup \ldots \cup C_{\nu i}, \quad i = 0, 1, \ldots, n.$$

Finally, we define the required system of events as follows:

$$A_1 \approx A'_1 \cup A'_2 \cup \ldots \cup A'_{c_1};$$

$$A_2 \approx A'_{c_1+1} \cup A'_{c_1+2} \cup \ldots \cup A'_{c_1+c_2};$$

. . . .

$$A_l \approx A'_{n-c_l+1} \cup \ldots \cup A'_{n-1} \cup A'_n;$$

$$A_0 \approx (\overline{A_1 \cup A_2 \cup \ldots \cup A_l}).$$

It is easy to see that this system satisfies conditions (a) and (b). It remains to verify condition (c). To this end, we first determine the number ν of events C_j. Let the number of significant components of the vector $\overline{q}$ be $h+\mu$. If the indices of these components are fixed, the significant components may be chosen in $(d-1)^{h+\mu}$ different ways. Now every choice of significant components also implies a choice of the remaining $f-\mu$ components. Since the latter may assume values 0 and d only, they may be chosen in $2^{f-\mu}$ different ways. The degree of freedom in choosing the indices of the significant components introduces yet another factor $C_r^{h+\mu}$. Thus the number of events $B_{\overline{q}}$ for vectors $\overline{q}$ with $h+\mu$ significant components is

$$(d-1)^{h+\mu} \cdot 2^{f-\mu} C_r^{h+\mu},$$

and since $\mu=0, 1, \ldots, f$, we have

$$\nu = \sum_{\mu=0}^{f} (d-1)^{h+\mu} \cdot 2^{f-\mu} C_r^{h+\mu}. \qquad (8)$$

Consequently,

$$\mu(A'_0) \not> (n-1) \sum_{\mu=0}^{f} (d-1)^{h+\mu} \cdot 2^{f-\mu} C_r^{h+\mu}. \qquad (9)$$

We now estimate the number of elementary events in C_0:

$$C_0 \approx (\overline{C_1 \cup C_2 \cup \ldots \cup C_\nu}).$$

It is clear that C_0 consists of the events $B_{\overline{q}}$ such that the number of significant components of $\overline{q}$ is at most $h-1$. We therefore reason as follows. Fix the indices and values of the nonsignificant components of $\overline{q}$. Let their indices be $i_1, i_2, \ldots, i_{h-\mu}$. Then the choice of values $q_{i_1}, q_{i_2}, \ldots, q_{i_{h-\mu}}$ for these components completely determines the number of elementary events contained in $B_{\overline{q}}$. Thus, for each fixed choice of nonsignificant components we have $(d-1)^{h-\mu}$ events $B_{\overline{q}}$, and the total number of elementary events in the latter is now given by

$$\sum_{(q_{i_1} \ldots q_{i_{h-\mu}})=(11\ldots1)}^{(d-1\ldots d-1)} C_d^{q_{i_1}} C_d^{q_{i_2}} \ldots C_d^{q_{h-\mu}} = (2^d-2)^{h-\mu}. \qquad (10)$$

But the number of choices of values for the nonsignificant components is $2^{f+\mu}$, and the number of possible choices for their indices is $C_r^{f+\mu}$. Consequently, the number of elementary events in the union of all events $B_{\overline{q}}$ such that $\overline{q}$ has exactly $h-\mu$ significant components is

$$(2^d-2)^{h-\mu} \cdot 2^{f+\mu} C_r^{f+\mu}.$$

Since $\mu=1, 2, \ldots, h$, the number of elementary events in C_0 is

$$\mu(C_0)=\sum_{\mu=1}^{h}(2^d-2)^{h-\mu}\cdot 2^{f+\mu}C_r^{h-\mu}. \qquad (11)$$

Using (9) and (11), we obtain

$$p(A_0)\not> \varrho^{dr}\left((n-1)\sum_{\mu=0}^{f}(d-1)^{h+\mu}\cdot 2^{f-\mu}C_r^{h+\mu}+\right.$$
$$\left.+\sum_{\mu=1}^{h}(2^d-2)^{h-\mu}\cdot 2^{f+\mu}C_r^{h-\mu}\right)=\varrho^{dr}2^r\left((n-1)\times\right.$$
$$\left.\times\sum_{\mu=0}^{f}C_r^{h+\mu}\left(\frac{d-1}{2}\right)^{h+\mu}+\sum_{\mu=1}^{h}C_r^{h-\mu}(2^{d-1}-1)^{h-\mu}\right)\not>$$
$$\not>(n-1)(\varrho^d(d+1))^r+2(2\varrho^d)^r(2^{d-1}-1)^{h-1}r^{h-1}.$$

Consequently,

$$p(A_0)\not>(n-1)(\varrho^d(d+1))^r+2(2\varrho^d)^r(2^{d-1}-1)^{h-1}r^{h-1}. \qquad (12)$$

This proves condition (c) and the proof is complete.

Theorem 3. Given a synthesis basis $\mathcal{B}=(\mathcal{A}_\neg, \mathcal{A}_\&, \mathcal{A}_\vee, \mathcal{A}_t, \mathcal{A}_1, \ldots, \mathcal{A}_r)$ *such that for some natural numbers* d, h, k

$$h\leqslant r; \qquad (13)$$
$$d^h(\varrho^d(d+1))^r+(2^{d-1}-1)^{h-1}r^{h-1}(2\varrho^d)^r<2^{-k-1}, \qquad (14)$$

where

$$\varrho=\max_i \omega((\mathbf{M}_i)_{\sigma_2}). \qquad (15)$$

Then, if

$$\forall i(\varkappa((\mathbf{M}_i)_{\sigma_2})=0) \qquad (16)$$

and $\pi_0=\left(\frac{c_1}{n},\ldots,\frac{c_l}{n}\right)$, $n\leqslant d^h$, *is any stochastic vector, one can construct a* P*-net* L *over* $\mathcal{B}$ *satisfying the following conditions:*

(1) L *realizes a stochastic vector* $\pi'_0=(p'_1, \ldots, p'_l)$ *such that*

$$\omega(\pi_0-\pi'_0)<2^{-k}. \qquad (17)$$

(2) For every i, $1\leqslant i\leqslant r$, *the element* $\mathcal{A}_i$ *is attached to exactly* d *vertices of* L.

(3) Any P*-net* L' *synthesized over a basis* $\mathcal{B}'=(\mathcal{A}_\neg, \mathcal{A}_\&, \mathcal{A}_\vee, \mathcal{A}_t, \mathcal{A}'_1, \ldots, \mathcal{A}'_r)$ *which is related to* L *realizes a stochastic vector* $\pi''_0=(p''_1, \ldots, p''_l)$ *with* $\omega(\pi_0-\pi''_0)<2^{-k}$, *provided*

$$\forall i(\varkappa((\mathbf{M}'_i)_{\sigma_2})=0) \tag{18}$$

and

$$\max_i \omega((\mathbf{M}'_i)_{\sigma_2}) \not> \varrho. \tag{19}$$

Proof. The P-net L is easily constructed by the method described in the last section. Using Lemma 2, one readily proves that the output words of the P-net L_1 of Figure 16* may be grouped into classes $D_1, \ldots, D_l$ such that the probability of L_1 going from any state η to a state η' corresponding to an output word $\alpha \in D_i$ is equal to $\frac{c_i}{n} + \varepsilon_i$, where $|\varepsilon_i| < 2^{-k-1}$.

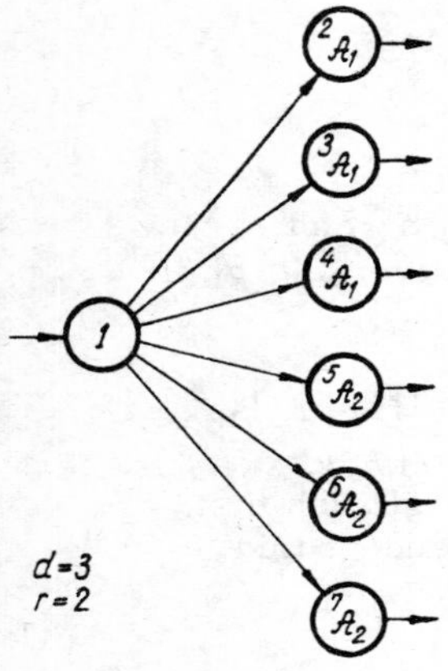

FIGURE 16.

Let the truth domains of the functions $f_1(x_1, x_2, \ldots, x_6)$ and $f_2(x_1, x_2, \ldots, x_6)$ be D_1 and D_2, respectively. Let L'_1 and L'_2 denote P-nets realizing f_1 and f_2, respectively. It is now clear that the P-net L shown in Figure 17 ($d=3$, $r=2$, $l=2$) answers our purposes.

It is not hard to see that slight modifications of conditions (14), (15) and (19) would enable us to prove Theorem 3 without conditions (16) and (18).

It is worth noting that, for fixed values of h and k, the admissible values of ϱ increase together with d and r; in other words, ϱ tends to 1 when d and r tend to infinity. This means that Theorem 1 is only a special case (in a certain sense) of Theorem 3. Nonetheless, a comparison of the two theorems in the light of the criteria we have been using reveals no real advantage of Theorem 3.

Certain advantages come to light only if we look in a different way at P-nets realizing stochastic vectors. To make our new position clear, we introduce the concept of a distribution generator.

Definition 2. Let L be a P-net with l output vertices and $[f, g]$ a pair of input words. Let π_f denote the first row of the matrix $\mathbf{L}_f$ and $\Omega' = (E', \mathfrak{B}')$ the field of first Markov r-trials induced by the Markov chain $[L_s, \pi_f, \mathbf{L}_g]$, where L_s is the ordered set of states of L. Let F_j, $j=1, 2, \ldots, l$, be the set of all states of L corresponding to an output word $\alpha \supseteq \alpha_1\alpha_2\ldots\alpha_l$, satisfying the conditions

$$\begin{cases} \alpha_j = 1; \\ \nu(\alpha) = 1. \end{cases}$$

We define a random variable $\mathfrak{P}_{ij}$, $i=1, 2, \ldots, r$, $j=1, 2, \ldots, l$ over Ω' by

$$\mathfrak{P}_{ij}(e') = \mathfrak{P}_{ij}(\delta\eta_{t_1}\,\delta\eta_{t_2}\,\delta\ldots\delta\eta_{t_r}\,\delta) = 1,$$

* In the figure, $d=3$, $r=2$.

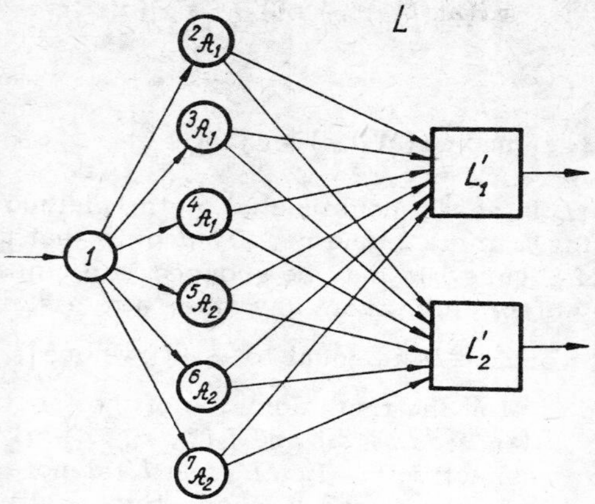

FIGURE 17.

if $\eta_{t_i} \in F_j$, and $\mathfrak{P}_{ij}(e')=0$ otherwise. We shall say that L generates the field of r-independent trials associated with $\Omega=(E, \mathfrak{B})$ relative to $[f, g]$, if

(1) $\mu(E)=l$;

(2) $\forall i \forall j (p(\{\mathfrak{P}_{ij}=1\})=\mathfrak{B}(e_j))$;

(3) $p(\{\mathfrak{P}_{1j_1}=1, \mathfrak{P}_{2j_2}=1, \ldots, \mathfrak{P}_{rj_r}=1\})=p(\{\mathfrak{P}_{1j_1}=1\})p(\{\mathfrak{P}_{2j_2}=1\})\ldots p(\{\mathfrak{P}_{rj_r}=1\})$.

Definition 3. A generator of the distribution $\pi_0=(p_1, p_2, \ldots, p_l)$ is specified through a P-net L and a pair of input words $[f, g]$ such that, for any r, L generates relative to $[f, g]$ the field of r-independent trials associated with a PF $\Omega=(E, \mathfrak{B})$ with the properties

(a) $\mu(E)=l$;

(b) $\forall j (\mathfrak{B}(e_j)=p_j)$.

It is quite easy to prove the following

Theorem 4. Let $L', L'_1, \ldots, L'_l$ be P*-nets with the properties:*

(1) The number n of output vertices of L' is the same as the number of input vertices of L'_i, $i=1, 2, \ldots, l$.

(2) Each P*-net L'_i has only one output vertex and the same depth $k\nu$.*

(3) L'_i, $i=1, 2, \ldots, l$, realizes a boolean function $f_i(x_1, x_2, \ldots, x_n)$ such that

$$(i\neq j)\supset(f_i(x_1, x_2, \ldots, x_n)f_j(x_1, x_2, \ldots, x_n)=0),$$
$$f_1 \cup f_2 \cup \ldots \cup f_l=1.$$

(4) Suppose the states of L' partitioned into classes F_i, $i=1, 2, \ldots, l$, such that $\eta\in F_i$ if and only if the output word α of L' corresponding to η is in the truth domain of the function $f_i(x_1, \ldots, x_n)$; there exists an input word g of length ν such that the matrix $\mathbf{L}'_g$ can be condensed with respect to the above classification of states.

Then the P*-net* L *obtained by identifying the* j*-th output vertex of* L' *with the* j*-th input vertex of* L', $j=1, \ldots, n$; $i=1, \ldots, l$, *generates the distribution* $\pi_0=(p_1, \ldots, p_l)$ *relative to the pair of input words* $[f, g]$ *if the condensed form of the matrix* $\mathbf{L}'_g$ *is a matrix* $\mathbf{P}=(p_{il})$ *all of whose rows coincide with* π_0.

We see that P-nets that realize certain stochastic vectors synchronously (these are in effect autonomous FSA of a special type) at the same time generate certain distributions. The numbers $l(f)$ and $l(g)$ are important characteristics of distribution generators; they are known respectively as the shift and the expansion. These characteristics may be used to compare different distribution generators as to their relative advantages and disadvantages.

NOTES

We have been considering only methods for the structural realization of stochastic vectors and distribution generators, which would seem to have little in common with the design of probability transformers. Despite this, the entire chapter is in actual fact concerned also with the synthesis of probability transformers. It is therefore quite legitimate to compare our methods with those of Gill, Sheng (1965), Kobchikov, Bukharaev, Makarevich and others. From the conceptual standpoint, the most interesting work is that of Gill (1962), Kobchikov (1967), and Makarevich (1969b, c, 1970). Gill's highly original paper is distinguished for the simplicity and flexibility with which he solves the problem. His method is simple as regards physical realization of the probability transformer; its flexibility reveals itself in that the distribution may be realized with a good tolerance even when the initial distribution is known only with poor tolerance. Of course, the desired tolerance is achieved at the cost of slowing down the process, i.e., increasing the expansion. In an attempt to increase the speed of the transformer, Gill (1963) proposes a different method of construction. However, since the use of the new method is associated with the condition that the initial distribution be known infinitely accurately, flexibility is lost. Moreover, Gill's derivation of an estimate for the tolerance as a function of the speed of the process is not effective and is thus suitable only for the field of rational numbers and inessential extensions thereof. Sheng's method (Sheng, 1965) suffers from the same drawback, although from the standpoint of physical implementation it offers interest for the regularity and simplicity of design of the transforming network. It is easily seen that the proof of Proposition 4 (see Chapter IV) is a constructive version of Gill's second idea. His estimate for the dependence of the accuracy on the expansion may be considerably sharpened, although in so doing one looses the regularity of the transforming network. This estimate is probably inferior to the classically proved estimate of Gill.

Kobchikov's paper presents an interesting technique for the design of controllable distribution generators, a subject of considerable importance.*

* The problem of designing controllable random distribution generators was considered previously by Bukharaev (1963b). He also outlined a technique for designing a generator as a relay contact network.

However, his method (like many others) presumes accurate knowledge of the initial distributions; in other words, it is not flexible. The design of controllable or (in a certain sense) universal probability transformers has been considered by Makarevich. A comparison of this method with that of Kobchikov should be undertaken with caution, since Makarevich considers asynchronous probability transformers.

Makarevich and Matevosyan (1970) proposed a flexible method for the synthesis of asynchronous distribution generators. The present author, working independently, developed another flexible method for the synthesis of asynchronous probability transformers (Lorenz, 1969d) (see also Proposition 5, Chap. VIII, and Theorem 2). Theorems 1 and 2 in this book are actually only specific realizations of the methods we developed there. It should be noted that in both the Makarevich-Matevosyan method and our method the accuracy of realization, or rather the stochastic vector (or distribution) being realized, is independent of the initial parameters or of the initial distribution $\pi_0 = (p, q)$ provided $0 < p < 1$. However, it is easy to see that no asynchronous method for synthesis of probability transformers (as is the case for synchronous methods) can possibly furnish one with an infinitely accurate realization of stochastic vectors unless it be possible to design absolutely stable basis elements or initial distributions. It is therefore of special interest to devise methods of structural synthesis enabling one to control the effect of unfavorable random variations in the initial distributions. Networks or P-nets possessing this property might be called stable. We note that Gill's first method (Gill, 1962) synthesizes stable probability transformers. This follows directly from the structure of the doubly stochastic matrices on which his method is based and from Theorem 2, Chap. VI. As for our synthesis methods (see Theorems 1 and 2), they are also flexible and may probably be rendered stable to some degree, to produce stable P-nets. In a few special cases this may be proved rigorously. At the same time, our methods yield P-nets whose action is more rapid than guaranteed in Gill's method. The merit of Bukharaev's papers (1963b, 1964d) is their formulation of the problem of developing methods for construction of suitably distributed random variables from given variables. This problem is clearly more general than the analogous problem for distributions.

Appendix

PRINCIPLES OF CONSTRUCTIVE MATHEMATICS

Before reading this appendix, the reader should gain some familiarity with the contents of Chapter I, since we shall be using some of the notation and concepts introduced there. Apart from this, an acquaintance with the material of Chapter I will help the reader to focus his attention on the appropriate points.

The constructive mathematician is interested only in constructive objects, their properties and relationships. The notion of a "constructive object" is not defined, but only explained. In so doing, we are fully aware that even after explanations are given many questions will remain unanswered. Moreover, explanations, whoever gives them, will always remain incomplete, invoking a feeling of dissatisfaction in the critically inclined. For example, just what does a constructivist understand by a "constructive object"? He is clearly not thinking of some "lump of material." The representatives of constructive mathematics often cite a child's construction toy as an intuitive illustration, where the basic parts (basic constructive objects) are used to construct new objects; the true essence of the concept is, however, different. Instead of constructive objects, it would be quite legitimate to speak of shapes (Lorenzen, 1955) formed in accordance with definite rules from other shapes, the latter being regarded as "raw material." These basic shapes may be specified through some list or alphabet and termed "letters" of the alphabet. For example, one might decide on the list of basic shapes in Figure 18, letting the rules for formation of new shapes be:

(a) each basic shape is a shape;

(b) if φ' is a shape and φ a basic shape, one obtains a new shape φ'' by bringing one of the straight sides of φ' (if such exist) into coincidence with a straight side of φ. Some shapes constructed from the basic shapes of our list are shown in Figure 19.

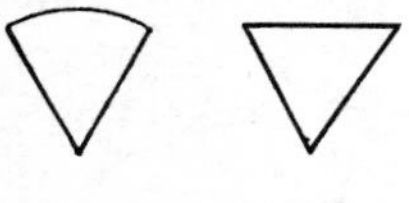

FIGURE 18.

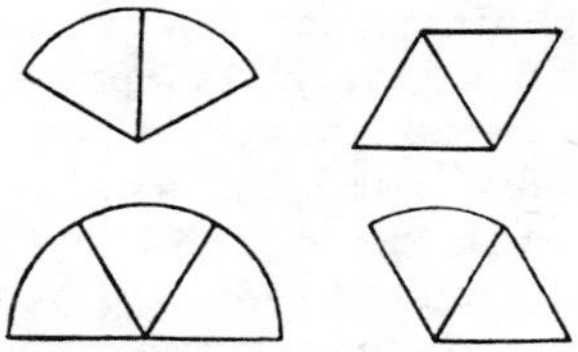

FIGURE 19.

A more widespread way of combining basic shapes is to arrange them in a sequence, i. e., to form finite linear sequences of shapes. The new shapes thus formed are known as words. For example, if our alphabet of basic shapes consists of the letters $\mathfrak{R}$, $\mathfrak{L}$, examples of constructive objects (shapes) are $\mathfrak{R}$, $\mathfrak{L}$, $\mathfrak{R}\mathfrak{R}$, $\mathfrak{L}\mathfrak{R}$, $\mathfrak{L}\mathfrak{L}$, etc. Frequently one allows the abstract object, defined as the empty place formed by removing a shape, to be a shape (word). Strictly speaking, one can avoid the use of the empty word — it is used only to simplify certain algorithms and operations.

Of course, such a treatment of the constructive object necessarily involves the abstraction of identification; otherwise one would have to deal with objects as absolute individuals. To illustrate: without using this abstraction, i. e., ignoring certain inessential features of some object such as the letter $\mathfrak{R}$, it would be impossible to speak of this letter as being equal to another occurrence $\mathfrak{R}$ of the "same" letter. Thus the relation of equality (graphical equality) between two letters and, correspondingly, two words, is an abstraction, which is not even fully defined. This is because there is no way of exhaustively listing all the features to be ignored in order that two words denoted by e and f, say, be considered graphically equal (the notation for this relation will be $e \,\overline{\underline{\circ}}\, f$). Clearly, graphical equality of two letters or words is largely a matter of convention. Thus, for example, in some contexts we shall view the letters A and **A** as graphically equal, in others we may distinguish between them. This situation has not as yet led to any misunderstandings (though it may in the future).

Another important abstraction made in constructive mathematics is the so-called abstraction of potential realizability. To elucidate: although within one life span a person can actually write down only a limite (finite) number of words, our reasoning envisages the possibility of achieving words of arbitrary finite length, and in arbitrary quantity. Of course, in so doing one is also disregarding the question of whether one can acquire an arbitrarily long tape or arbitrarily large board on which to write the required words. For example, if we define the natural numbers as words of the form $0, 0|, 0||, \ldots$ over the alphabet A_0, the abstraction of potential realizability gives us the right to speak of arbitrarily long words of this type or arbitrarily large natural numbers. By contrast, constructive mathematics rejects the abstraction of actual infinity, according to which one may, for example, treat the natural number sequence as a complete whole, as a well-defined "infinite" object. Along with actual infinity, one is obviously also rejecting any argument and/or definition based upon it.

In the theory of algorithms, the abstraction of potential realizability is also inherent in the fact that of the operations prescribed by an algorithm scheme only a limited number may actually be performed, and moreover errors may be made in the process. One always reasons as if the elementary operations may be implemented in the required sequence without error and in any number.

Despite the fact that the constructivists reject the abstraction of actual infinity, they are able to define and handle infinite series, infinite sequences and infinite sets, at the same time avoiding the paradoxes and anomalies encountered in classical mathematics. In constructive mathematics, sequences and series of numbers are defined, as are functions, to be algorithms with suitable properties. For example, the constructivist defines a sequence of CRNs as an algorithm, applicable to any natural number, which produces a CRN. Since any normal algorithm may be encoded as a suitable word, number sequences and functions are in the final analysis constructive objects, with which one can work as freely as with other words. For example, one can define a sequence of such words, i. e., a sequence of sequences or sequence of functions.

Sets of words, including numbers, are also defined as constructive objects, but their true nature is another matter. They are in effect notations for one-parameter conditions or properties, in terms of rigorous linguistic tools. Thus, for the constructive mathematician the set of prime numbers is not an imaginary infinite object consisting of all natural numbers m divisible only by themselves and by 1, but only a condition, e. g.,

$$\forall_{1<n<m} n \ (\text{res} \frac{m}{n} \neq 0).$$

Consequently, an infinite set is for the constructivist not the "finished product," i. e., the totality of words satisfying a condition, but just a way of writing down the condition.

Constructive mathematics allots a major role to what are known as generable sets. These are defined by systems of rules with the following structure:

(1) The initial constructive objects are fixed, i. e., a list of basic objects is specified.

(2) Rules for making new objects out of old are stipulated.

Sometimes these two conditions are supplemented by a third, which specifies a condition imposed on the constructed objects, which might be called a restricting condition. It is readily understood that generable sets are defined by means of a suitable fundamental induction (Kleene, 1952). Any generable set defined without a restricting condition may be represented by some condition of the form

$$\mathfrak{A}(x) > 0, \qquad (1)$$

where $\mathfrak{A}$ is a normal algorithm transforming a word x into a CRN. An exception is the set of all words over the alphabet. One cannot say that this set is defined by a condition of type (1), since the condition presupposes the existence of the set of words over the alphabet. Thus, generable sets play a fundamental role in constructive mathematics.

Another aspect in which constructive mathematics departs in an essential manner from classical mathematics is in the interpretation of logical connectives and quantifiers. A disjunction of two statements P and Q, $P \vee Q$, is deemed true only if one has a procedure verifying that one of the statements is true. The constructivist cannot regard the statement $P \vee Q$ as true if he can only derive a contradiction by assuming that $\neg (P \vee Q)$ is true. For example, given any CRN d one readily derives a contradiction from the assumption that the statement $\neg((d=0) \vee (d \neq 0))$ is true. But since these arguments alone furnish no procedure verifying the truth of one of the statements $d=0$, $d \neq 0$, they are no proof that the statement $(d=0) \vee (d \neq 0)$ is true. For any given d, then, the truth of $(d=0) \vee (d \neq 0)$ must be proved by other means.

The implication $P \supset Q$ of two statements P, Q is usually understood in constructive mathematics as deducibility of Q from P. Lately, however, this connective has been interpreted in a more subtle way, especially in the work of Shanin.

As for the negation of a statement P, i. e., $\neg P$, in constructive mathematics this means that a falsehood may be deduced from the assumption that P is true. Naturally, this interpretation of negation demands that we have at our disposal statements whose falsehood either has been stipulated as a convention or represents an empirical fact. This role is usually given to statements such as $0 \mathrel{\overline{\overset{\circ}{=}}} |$, $| \mathrel{\overline{\overset{\circ}{=}}} ||$, $|| \mathrel{\overline{\overset{\circ}{=}}} |||$, etc. Obviously, the truth of the statement $\neg (| \mathrel{\overline{\overset{\circ}{=}}} ||)$ is a fact of our experience only because two letters are viewed as graphically equal or distinct by dint of the abstraction of identification. Shanin goes even farther in this respect, demanding that negated statements be handled directly, that is to say, without reference to deducibility of some falsehood.

The constructivist interpretation of the existential quantifier $\exists$ is also not standard. The statement $\exists xP(x)$ is deemed true only if one has a procedure producing a word e for which the statement $P(e)$ is true. Thus the passage from $\neg\neg(\exists xP(x))$ to $\exists xP(x)$, permissible in classical mathematics, is prohibited in constructive mathematics. It is legitimate only in case the statement $(P(x)) \vee (\neg P(x))$ is true for every word x. This logical principle, first introduced by Markov, is known as the principle of constructive choice (or the Leningrad principle). The intuitionists reject it and some of them have even tried (without success) to discredit it. The main argument of the intuitionists is this: according to the principle of constructive choice, the existential quantifier may be introduced, ostensibly, without providing a procedure for construction of an object (word) e such that $P(e)$ is true. However, it is easy to see that the required procedure is in fact implicit in the logical principle itself. Indeed, the principle of constructive choice, written out, say, as a scheme of formal interference:

$$\frac{\forall x(P(x) \vee \neg P(x)), \neg\neg\exists xP(x)}{\exists xP(x)}$$

shows that one can determine the required value of x by successively trying new words.* This procedure must end sooner or later, for otherwise it would follow that the statement $\neg\exists xP(x)$ is true, and this contradicts the premise $\neg\neg\exists xP(x)$. Consequently, after a certain number of trials one necessarily "runs across" a word e satisfying the condition $P(x)$. For the intuitionists, however, this procedure is unacceptable: in their view it is ineffective, since the number of steps cannot be bounded in advance. It is also noteworthy that the constructivists utilize the principle of constructive choice to decide whether or not an algorithm is applicable to a given word. Thus, the intuitionists, rejecting as they do the principle of constructive choice, thereby reject the entire theory of algorithms as an acceptable rigorous explication of the concept "effectively computable function." On the contrary, they are agreed in viewing an algorithm $\mathfrak{A}$ as applicable

* It is assumed that any word over the alphabet may indeed by reached (potentially) in this process.

to a word e over an alphabet B only if one can indicate an effectively computable function $\varphi(x)$ (in the intuitionist sense) satisfying the condition: for any word e, the algorithm $\mathfrak{A}$ halts, i. e., yields a result $\mathfrak{A}(e)$, in $m_e \leqslant \varphi(e)$ steps.

As to the universal quantifier $\forall$, the representatives of Markov constructivism adhere to the following conventions: the statement $\forall xP(x)$ is true (proved) if one can indicate a general method to prove the truth of $P(x)$, suitable for any prescribed value of the variable x. In other words, the proof of a statement of type $\forall xP(x)$ is so to speak a parametric proof, which becomes a proof of $P(e)$ for any value e of x. Consequently, there is no application here of the abstraction of actual infinity and we are still within the bounds of the abstraction of potential realizability.

The general rules of inference that the constructivists use to prove theorems may be codified if so desired. The result is then an intuitionistic variant of Gentzen's predicate calculus (Gentzen, 1934). One additional rule is the principle of constructive choice. In actual fact, however, the situation is not quite as described. Like the intuitionists, the constructivists do not acknowledge the sovereignty of logic. Logic is always referred to a certain language, to a certain class of statements phrased in that language. This unity of logic and language is most clearly exemplified in Markov (1968). Logical rules are introduced there relative to a rigidly fixed mathematical language, i. e., they refer not to statements as such but rather to statements in the language. In such a context, it may well happen that logical principles which are usually treated as independent rules are in fact derived rules. This is precisely the case as regards the principle of constructive choice.

It is often asserted that the constructivists reject the law of the excluded middle. This assertion should be understood in the light of the foregoing discussion: everything depends on the language under consideration and on the corresponding interpretation of the logical connectives. If one is considering an extremely simple language, or is interested in very simple statements such as statements concerning natural numbers of the type $n=m$, $n>m$, $n<m$ and their negations, deciding to use in addition to these only their disjunctions and conjunctions, the law of the excluded middle is eminently acceptable to the constructivist. The situation changes as the language becomes richer; this is clearly demonstrated in many parts of our book.

Finally, the constructivists treat conjunctions $P\&Q$ as in classical mathematics: $P\&Q$ is true if and only if both P and Q are true statements.

BIBLIOGRAPHY

Agafonov, V.N. Enumerability on stochastic machines and hypersimple sets. — In: Tezisy dokladov na Vsesoyuznom simpoziume po veroyatnostnym avtomatam. Kazan', 1969a. (Russian)

Agafonov, V.N. On algorithms, frequency and randomness. — Author's summary of candidate dissertation. Novosibirsk, 1969b. (Russian)

Agafonov, V.N. More about stochastic machines. — Tezisy Dokladov V Vsesoyuznogo simpoziuma po kibernetike. Tbilisi, 1970. (Russian)

Arbib, M.A. Theories of Abstract Automata. — Englewood Cliffs, N.Y., Prentice-Hall, 1969.

Arbib, M.A. and M. Blum. Machine dependence of degrees of difficulty. — Proc. Amer. Math. Soc. **16**, No.3 (1965).

Areshyan, G.L. and G.B. Marandzhyan. On some questions of stochastic automata. — In: Matematicheskie voprosy kibernetiki i vychislitel'noi tekhniki, Vol.2. Erevan, Izdatel'stvo Akad. Nauk Arm. SSR, 1964. (Russian)

Ashby, W.R. An Introduction to Cybernetics. — London, Chapman and Hall, 1956.

Atkinson, R.C., G.H. Bower and E.J. Crothers. An Introduction to Mathematical Learning Theory. — New York, Wiley, 1965.

Bacon, G.C. Minimal-state stochastic finite-state systems. — IEEE Trans. Circuit Theory CT-11 (1964a).

Bacon, G.C. The decomposition of stochastic automata. — Information and Control 7 (1964b).

Barzdin', Ya. M. Complexity of recognition of symmetry on Turing machines. — Problemy Kibernetiki 15 (1965). (Russian)

Barzdin', Ya.M. Complexity of programs that decide whether natural numbers not exceeding x belong to a recursively enumerable set. — Dokl. Akad. Nauk SSSR **182**, No.6 (1968). (Russian)

Barzdin', Ya.M. On computability in stochastic machines. — Dokl. Akad. Nauk SSSR **189**, No.4 (1969). (Russian)

Barzdin', Ya.M. On identification of automata. — Problemy Kibernetiki **21** (1969). (Russian)

Barzdin', Ya.M. On the frequency solution of algorithmically unsolvable general problems. — Dokl. Akad. Nauk SSSR **191**, No.5 (1970). (Russian)

Berge, C. Théorie des graphes et ses applications. — Paris, Dunod, 1958.

Blum, M. A machine-independent theory of the complexity of recursive functions. — J. Assoc. Comput. Mach. **14**, No.2 (1967a).

Blum, M. On the size of machines. — Information and Control **11**, No.3 (1967b).

Boichenko, V.M. and V.S. Gladkii. Simulation of operations over stochastic matrices. — Izv. Akad. Nauk SSSR Tekhn. Kibernet., No.5 (1965a). (Russian)

Boichenko, V.M. and V.S. Gladkii. Stochastic simulation of the operations of multiplication and raising to a power of nonstochastic matrices. — Izv. Akad. Nauk SSSR, Tekhn. Kibernet., No.6 (1965b). (Russian)

Büchi, J.R. Weak second-order arithmetic and finite automata. — Z. Math. Logik Grundlagen Math. 6 (1960).

Büchi, J.R. Mathematische Theorie des Verhaltens endlicher Automaten. — Z. Angew. Math. Mech. 42 (1962).

Büchi, J.R. Regular canonical systems. — Arch. Math. Logik Grundlagenforsch. 6 (1963).

Büchi, J.R. Algebraic theory of feedback in discrete systems, I. — In: Automata Theory. New York, Academic Press, 1966.

Bukharaev, R.G. On the simulation of probability distributions. — Veroyatnostnye metody i kibernetika. Uchen. Zap. Kazan. Gos. Univ. 123, No.2 (1963a). (Russian)

Bukharaev, R.G. On controllable random-variable generators. — Veroyatnostnye metody i kibernetika. Uchen. Zap. Kazan. Gos. Univ. **123**, No.2 (1963b). (Russian)

Bukharaev, R.G. Questions of representability of extended communication channels and stochastic sequences in a stochastic automaton. — In: Itogovaya nauchnaya konferentsiya Kazanskogo gosudarstvennogo universiteta za 1963 g. Kazan', 1964a. (Russian)

Bukharaev, R.G. Equivalence criteria and equivalent transformers of stochastic Mealy automata. — In: Itogovaya nauchnaya konferentsiya Kazanskogo gosudarstvennogo universiteta za 1963 g. Kazan', 1964b. (Russian)

Bukharaev, R.G. Some equivalences in the theory of stochastic automata. — Veroyatnostnye metody i kibernetika. Uchen. Zap. Kazan. Gos. Univ. 124, No.2 (1964c). (Russian)

Bukharaev, R.G. On a random variable implying a certain commonly used class of random variables. — Veroyatnostnye metody i kibernetika. Uchen. Zap. Kazan. Gos. Univ. 124, No.2 (1964d). (Russian)

Bukharaev, R.G. Criteria for representation of events in finite stochastic automata. — Dokl. Akad. Nauk SSSR 164, No.2 (1965). (Russian)

Bukharaev, R.G. Automatic transformation of stochastic sequences. — Veroyatnostnye metody i kibernetika. Uchen. Zap. Kazan. Gos. Univ. 125, No.6 (1966a). (Russian)

Bukharaev, R.G. Two corrections to my paper "Some equivalences in the theory of stochastic automata." — Veroyatnostnye metody i kibernetika. Uchen. Zap. Kazan. Gos. Univ. 125, No.6 (1966b). (Russian)

Bukharaev, R.G. Toward a criterion for representation of events in finite stochastic automata. — Veroyatnostnye metody i kibernetika. Uchen. Zap. Kazan. Gos. Univ. 127, No.5 (1967a). (Russian)

Bukharaev, R.G. On representation of events in stochastic automata. — Veroyatnostnye metody i kibernetika. Uchen. Zap. Kazan. Gos. Univ. 127, No.3 (1967b). (Russian)

Bukharaev, R.G. The theory of stochastic automata. — Kibernetika, No.2 (1968a). (Russian)

Bukharaev, R.G. On sets of sequences of characteristic values of stochastic matrices. — Dokl. Akad. Nauk SSSR 178, No.2 (1968b). (Russian)

Bukharaev, R.G. Criteria for representation of events in finite stochastic automata. — Kibernetika, No.1 (1969). (Russian)

Bukharaev, R.G. Stochastic Automata. — Kazan', Izdatel'stvo Kazanskogo Gosudarstvennogo Universiteta, 1970. (Russian)

Bush, R.R. and F. Mosteller. Stochastic Models for Learning. — New York, Wiley, 1955.

Caldwell, S.H. Switching Circuits and Logical Design. — New York, Wiley, 1958.

Carlyle, J.W. Reduced forms for stochastic sequential machines. — J. Math. Anal. Appl.7, No.2 (1963).

Carlyle, J.W. On the extremal probability structure of finite-state channels. — Information and Control 7, No.3 (1964).

Carlyle, J.W. State-calculable stochastic sequential machines, equivalences and events. — In: IEEE 6th Ann. Symp. on Switching Circuit Theory and Logical Design. Ann Arbor, Mich. 1965.

Chentsov, V.M. Synthesis of stochastic automata. — In: Problemy sinteza tsifrovykh avtomatov. Moscow, Nauka, 1967. (Russian)

Chentsov, V.M. On a method for synthesis of an autonomous stochastic automaton. — Kibernetika, No.3 (1968a). (Russian)

Chentsov, V.M. Study of the behavior of stochastic automata with variable structure. — In: Informatsionnye seti i kommutatsiya. Moscow, Nauka, 1968b. (Russian)

Chirkov, M.K. Toward analysis of stochastic automata. — Vychislitel'naya Tekhnika i Voprosy Programmirovaniya. No.4 (1965). (Russian)

Chirkov, M.K. Stochastic automata and stochastic mappings. — Diskretnyi Analiz, No.7 (1966). (Russian)

Cleave, J.P. The synthesis of finite-state homogeneous Markov chains. — Cybernetica, No.5 (1962).

Davis, A.S. Markov chains as random input automata. — Amer. Math. Monthly 68, No.3 (1961).

Dijkman, J.G. On Markov chains and intuitionism. — Nederl. Akad. Wetensch. Proc. Ser. A64, No.3 (Indag. Math. 23, No.3) (1961).

Dijkman, J.G. On Markov chains and intuitionism. II. Discrete state-space and continuous parameter. — Nederl. Akad. Wetensch. Proc. Ser. A66, No.3 (Indag. Math. 25, No.3) (1963).

Dijkman, J.G. On Markov chains and intuitionism, III. Note on continuous functions with an application to Markov chains. — Nederl. Akad. Wetensch. Proc. Ser. A 67, No.3 (Indag. Math. 26, No.3) (1964a).

Dijkman, J.G. On Markov chains and intuitionism, IV. — Nederl. Akad. Wetensch. Proc. Ser. A 67, No.5 (Indag Math. 26, No.5) (1964b).

Dijkman, J.G. Probability theory and intuitionsim: discrete state-space. — Compositio Math. 17, No.1 (1965).

Doob, J.L. Stochastic Processes. — New York, Wiley, 1953.

Even, S. Rational numbers and regular events. — IEEE Trans. Electronic Computers EC-13 (1964).

Even, S. Comments on the minimization of stochastic machines. — IEEE Trans. Electronic Computers EC-14 (1965).

Feichtinger, G. Stochastische Automaten als Grundlage linearer Lernmodelle. — Statistische Hefte 10, No.1 (1969a).

Feichtinger, G. Ein Markoffsches Lernmodell für Zwei-Personen-Spiele. — Elektron. Datenverarbeitung, No.7 (1969b).

Feichtinger, G. Lernprozesse in stochastischen Automaten. — Berlin, Springer, 1970a.

Feichtinger, G. Gekoppelte stochastische Automaten und sequentielle Zwei-Personen-Spiele. — Unternehmensforsch. (1970b).

Feichtinger, G. "Wahrscheinlichkeitslernen" in der statistichen Lerntheorie. — Metrika (1970c).

Feller, W. An Introduction to Probability Theory and its Applications, Vol.I. — New York, Wiley, 1950.

Gantmakher, F.R. Theory of Matrices. — Moscow, Nauka, 1967. (Russian)

Gavrilov, M.A. Theory of Relay-Contact Networks. — Moscow, Izdatel'stvo Akad. Nauk SSSR, 1950. (Russian)

Gentzen, G. Untersuchungen über das logische Schliessen, I, II. — Math. Z. 39 (1934).

Gill, A. Synthesis of probability transformers. — J. Franklin Inst. 274, No.1 (1962).

Gill, A. On a weight distribution problem with application to the design of stochastic generators. — J. Assoc. Comput. Mach. 10, No.1 (1963).

Gill, A. Analysis and synthesis of stable linear sequential circuits. — J. Assoc. Comput. Mach. 12, No.1 (1965).

Gladkii, V.S. Matrix inversion using stochastic automata. — In: Veroyatnostnye avtomaty i ikh primenenie. Riga, Zinatne. 1971. (Russian)

Glova, V.I. On synthesis of devices which are stochastically multistable in regard to appearance of binary symbols. — In: Veroyatnostnye avtomaty i ikh primenenie. Riga, Zinatne, 1971. (Russian)

Glushkov, V.M. Abstract theory of automata. — Uspekhi Mat. Nauk 18 (1961). (Russian)

Glushkov, V.M. Synthesis of Discrete Automata. — Moscow, Fizmatgiz, 1962. (Russian)

Glushkov, V.M., V.A. Kovalevskii and V.S. Mikhalevich. On the reliability of discrete automata. — In: Trudy VI Vsesoyuznogo soveshchaniya po teorii veroyatnostei i matematicheskoi statistike. Vil'nyus, 1962. (Russian)

Goodstein, R.L. Function theory in an axiom-free equation calculus. — Proc. London Math. Soc. Ser. 2 48 (1945).

Goodstein, R.L. Recursive Analysis. — Amsterdam, 1961.

Grenander, U. Can we Look Inside an Unreliable Automaton? — London, Wiley, 1966.

Harary, F., R.Z. Norman and D. Cartwright. Structural Models: an Introduction to the Theory of Directed Graphs. — New York, Wiley, 1965.

Heyting, A. Intuitionism. — Amsterdam, North-Holland, 1956.

Huffman, D.A. The synthesis of sequential switching circuits. — J. Franklin Inst. 257, Nos. 3, 4 (1954).

Huffman, D.A. The design and use of hazard-free switching networks. — J. Assoc. Comput. Mach. 4, No.1 (1957).

Kandelaki, N.P. and T.N. Tsertsvadze. On the behavior of some classes of stochastic automata in random media. — Avtomatika i Telemekh., No.6 (1966). (Russian)

Kashyap, R.L. Optimization of stochastic finite-state systems. — IEEE Trans. Automatic Control AC-11, No.4 (1966).

Kemeny, J.G. and J.L. Snell. Finite Markov Chains. — Princeton, N.J., Van Nostrand, 1959.

Khoury, D.J. Synchronizing sequences for probabilistic automata. — Stud. Appl. Math. 49, No.1 (1970).

Kleene, S.C. Introduction to Metamathematics. — Princeton, N.J., Van Nostrand, 1952.

Kleene, S.C. Representation of events in nerve nets and finite automata. — In: Automata Studies (Ann. Math. Studies, No.34), Princeton University Press, 1956.

Kleene, S.C. and R.E. Vesley. The Foundations of Intuitionistic Mathematics, Especially in Relation to Recursive Functions. — Amsterdam, North-Holland, 1965.

Knast, R. O pewnej możliwości syntezy strukturalnej automatu probabilistycznego. — Prace komisji budowy maszyn i elektrotechniki (Poznánskie towarzystwo przyjaciół nauk) 1, No.5 (1967).

Knast, R. Representability of nonregular languages in finite probabilistic automata. — Information and Control 16, No.3 (1970).

Kobchikov, A.V. A method for producing binary symbols with a prescribed probability of appearance. — Izv. Akad. Nauk SSSR, Tekhn. Kibernet., No.6 (1967). (Russian)

Kobrinskii, N.E. and B.A. Trakhtenbrot. On the construction of a general theory of logic networks. — In: Logicheskie issledovaniya. Moscow, Izdatel'stvo Akad. Nauk SSSR, 1959. (Russian)

Kochkarev, B.S. On the problem of stability of stochastic automata. — Veroyatnostnye metody i kibernetika. Uchen. Zap. Kazan. Gos. Univ. 127, No.3 (1967). (Russian)

Kochkarev, B.S. On the stability of stochastic automata. — Kibernetika, No.2 (1968a). (Russian)

Kochkarev, B.S. On partial stability of stochastic automata. — Dokl. Akad. Nauk SSSR 182, No.5 (1968b). (Russian)

Kochkarev, B.S. On testing the validity of a certain sufficient condition for stability of stochastic automata. — Kibernetika, No.4 (1969). (Russian)

Kosovskii, N.K. Necessary and sufficient conditions for Specker properties of a probability space. — Zap. nauchnykh seminarov Leningradskogo otdeleniya matematicheskogo instituta im. V.A.Steklova Akad. Nauk SSSR 16. Issledovaniya po konstruktivnoi matematike i matematicheskoi logike, II. Leningrad, Nauka, 1969. (Russian)

Kosovskii, N.K. FR-constructs on probability spaces. — Ibid.

Kosovskii, N.K. Laws of large numbers in constructive probability theory. — Ibid.

Kovalenko, I.K. Note on complexity of representation of events in stochastic and deterministic automata. — Kibernetika, No.2 (1965). (Russian)

Küstner, H. Analyse und Synthese stochastischer Automaten. — Elektronische Informationsverarbeitung und Kybernetik 5, No. 4/5 (1969).

Lazarev, V.G. and V.M. Chentsov. On minimizing the number of internal states of a stochastic automaton. — In: Sintez diskretnykh avtomatov upravlyayushchikh ustroistv. Moscow, Nauka, 1968. (Russian)

Leeuw, K. de, E.F. Moore, C.E., Shannon and N. Shapiro. Computability by probabilistic machines. — In: Automata Studies (Ann. Math. Studies, No.34). Princeton University Press, 1956.

Levin, V.I. Probability analysis of finite automata and their reliability. — Izv. Akad. Nauk SSSR, Tekhn. Kibernet., No.5 (1965). (Russian)

Levin, V.I. Determination of the characteristics of stochastic automata with feedback. — Izv. Akad. Nauk SSSR, Tekhn. Kibernet. No.3 (1966). (Russian)

Levin, V.I. An operational method for the study of stochastic automata. — Avtomat. i Telemekh., No.1 (1967). (Russian)

Levin, V.I. Analysis of the reliability of inhomogeneous Markov automata. — In: Nadezhnost' i effektivnost' diskretnykh sistem. Riga, Zinatne, 1968. (Russian)

Levin, V.I. Multiplication of weakly perturbed matrices. — Zh. Vychisl. Mat. i Mat. Fiz. 9, No.4 (1969a). (Russian)

Levin, V.I. Probability Analysis of Unreliable Elements. — Riga, Zinatne, 1969b. (Russian)

Loève, M. Probability Theory. — Princeton, N.J., Van Nostrand, 1960.

Lofgren, L. On the concept of self-repair, the limits of redundancy of circuits, and the capacity of a computation channel. — In: Trudy mezhdunarodnogo simpoziuma "Teoriya konechnykh i veroyatnostnykh avtomatov" (International Symposium on the Theory of Finite and Probabilistic Automata). Moscow, Nauka, 1965. (Russian)

Lorenz, A.A. Some questions in the constructive theory of finite stochastic automata. — Avtomat. i Vychisl. Tekhnika, No.5 (1967). (Russian)

Lorenz, A.A. Generalized quasidefinite finite stochastic automata and some algorithmic problems. — Avtomat. i Vychisl. Tekhnika, No.5 (1968a). (Russian)

Lorenz, A.A. Some questions in the constructive theory of finite stochastic automata. — Z. Math. Logik Grundlagen Math.14, No.5 (1968b). (Russian)

Lorenz, A.A. Questions of the reducibility of finite stochastic automata. — Avtomat. i Vychsl. Tekhn., No.1 (1969a). (Russian)

Lorenz, A.A. Saving of states in finite stochastic automata. — Avtomat. i Vychisl. Tekhn., No.2 (1969b). (Russian)

Lorenz, A.A. Elements of constructive probability theory. — Z. Math. Logik Grunlagen Math. 15 (1969c). (Russian)

Lorenz, A.A. Synthesis of stable finite stochastic automata. — Avtomat. i Vychisl. Tekhn., No. 4 (1969d). (Russian)

Lorenz, A.A. Problems of the constructive theory of stochastic automata. — In: Veroyatnostnye avtomaty i ikh primenenie. Riga, Zinatne, 1971. (Russian)

Lorenzen, P. Einführung in die operative Logik und Mathematik. — Berlin, Springer, 1955.

Makarevich, L.V. The problem of completeness in the structural theory of finite stochastic automata.— In: Konferentsiya po teorii avtomatov i iskusstvennomu myshleniyu. Annotatsiya dokladov (dopolnenie). Tashkent, Fan, 1968. (Russian)

Makarevich, L.V. On attainability in stochastic automata. — Soobshch. Akad. Nauk Gruz. SSR 53, No.2 (1969a). (Russian)

Makarevich, L.V. On a general approach to the structural theory of stochastic automata. — In: Materialy konferentsii molodykh uchenykh. Sbornik trudov Instituta kibernetiki Akad. Nauk Gruz. SSR. Tbilisi, Izdatel'stvo Akad. Nauk Gruz. SSR, 1969b. (Russian)

Makarevich, L.V. On realizability of stochastic operators in logical nets. — Diskretnyi Analiz, No.15 (1969c). (Russian)

Makarevich, L.V. On the problems of complexity and completeness in the theory of stochastic automata. — Author's summary of Candidate Dissertation. Novosibirsk, 1970. (Russian)

Makarevich, L.V. On a general approach to the structural theory of stochastic automata. — In: Veroyatnostnye avtomaty i ikh primenenie. Riga, Zinatne, 1971. (Russian)

Makarevich, L.V. and A.Kh. Giorgadze. On the question of the structural theory of stochastic automata. — Soobshch. Akad. Nauk Gruz. SSR 50. No.1 (1968). (Russian)

Makarevich, L.V. and A.A. Matevosyan. Transformation of random sequences in automata. — Avtomat. i Vychisl. Tekhn., No.5 (1970). (Russian)

Makarov, S.V. On the reliability of sequential circuits with small memory. — In: Vychislitel'nye sistemy, No.5. Novosibirsk, 1962. (Russian)

Makarov, S.V. On realization of stochastic matrices by finite automata. — In: Vychislitel'nye sistemy, No.9 Novosibirsk, 1963. (Russian)

Makarov, S.V. Probability calculations for sequential circuits. — In Vychislitel'nye sistemy, No.13. Novosibirsk, 1964. (Russian)

Marandzhyan, G.B. On a method for description of Markov automata. — Voprosy radioelektroniki, Ser. XII, Obshchetekhnicheskaya, No.25 (1965). (Russian)

Markov, A.A. Theory of algorithms. — Trudy Mat. Inst. Steklov 42 (1954). [English translation: Jerusalem, Israel Program for Scientific Translation, 1962]

Markov, A.A. On constructive functions. — Trudy Mat. Inst. Steklov 52 (1958). (Russian)

Markov, A.A. On constructive mathematics. — Trudy Mat. Inst. Steklov 67 (1962). (Russian)

Markov, A.A. On some algorithms connected with systems of words. — Izv. Akad. Nauk SSSR Ser. Mat. 27 (1963). (Russian)

Markov, A.A. On normal algorithms computing boolean functions. — Dokl. Akad. Nauk SSSR 157, No.2 (1964). (Russian)

Markov, A.A. On normal algorithms computing boolean functions. — Izv. Akad. Nauk SSSR, Ser. Mat.31, No.1 (1967). (Russian)

Markov, A.A. An approach to constructive mathematical logic. — In: Logic, Methodology and Philosophy of Sciences, III. Amsterdam, North-Holland, 1968.

McNaughton, R. Sequential circuits, Part B. Behavioral properties. — IEEE Trans. Circuit Theory CT-11, No.1 (1964).

McNaughton, R. Testing and generating infinite sequences by a finite automaton. — Information and Control 9 (1966).

Medvedev, Yu.T. On a class of events representable in a finite automaton. — In: Avtomaty. Moscow, Innostrannaya Literatura, 1956. (Russian) [Supplement to Russian translation of "Automata Studies," Princeton University Press, 1956]

Metra, I.A. Remarks on the minimal implicant for a stochastic matrix. — Avtomat. i Vychisl. Tekhn., No.5 (1970). (Russian)

Metra, I.A. Comparison of the number of states of stochastic and deterministic automata representing given events. — Avtomat. i Vychisl. Tekhn., No.5 (1971a). (Russian)

Metra, I.A. A stochastic counter. — In: Veroyatnostnye avtomaty i ikh primenenie. Riga, Zinatne, 1971b. (Russian)

Metra, I.A. and A.A. Smilgais. On some possibilities of representation of nonregular events by stochastic automata. — In: Latviiskii matematicheskii ezhegodnik, No.3. Riga, Zinatne, 1968a. (Russian)

Metra, I.A. and A.A. Smilgais. On the definiteness and regularity of events represented by stochastic automata. — Avtomat. i Vychisl. Tekhn., No.4 (1968b). (Russian)

Moisil, G.C. Teoria algebrică a mechanismelor automate. — Bucharest, Editura tehnică, 1959.

Moore, E.F. Gedanken-experiments on sequential machines. — In: Automata Studies (Ann. Math. Studies, No.34), Princeton University Press, 1956.

Moore, E.F. and C.E. Shannon. Reliable circuits using less reliable relays. — J. Franklin Inst. 3, No.4 (1956).

Muchnik, A.A. and S.G. Gindikin. On completeness of unreliable elements realizing boolean functions.— Dokl. Akad. Nauk SSSR 144, No.5 (1962). (Russian)

Murrey, F.J. Mechanisms and robots. — J. Assoc. Comput. Mach. 2, No.2 (1955).

Nagornyi, N.M. On sharpening the reduction theorem of the theory of normal algorithms. — Dokl. Akad. Nauk SSSR 90, No.3 (1953). (Russian)

Nawrotzki, K. Eine Bemerkung zur Reduktion stochastischer Automaten. — Elektronische Informationsverarbeitung und Kybernetik 2, No.3 (1966).

Neumann, J. von. Probabilistic logic and the synthesis of reliable organisms from unreliable components.— In: Automata Studies (Ann. Mat. Studies, No.34). Princeton University Press, 1956.

Nieh, T.T. and J.W. Carlyle. On the deterministic realization of stochastic finite-state machines. — In: Proc. 2nd Annual Princeton Conference on Information Science and Systems. 1968.

Onicescu, O. and S. Guiasu. Finite abstract random automata. — Z. Wahrscheinlichkeitstheorie und Verw. Gebiete 3 (1965)

Ore, O. Theory of Graphs. — New York, American Mathematical Society, 1962.

Ott, E H. Theory and Application of Stochastic Sequential Machines. — Research Report, Sperry Rand Research Center, Sudbury, Mass. 1966a.

Ott, E.H. Reconsider the state minimization problem for stochastic finite-state systems. — IEEE Conf. Rec. 7th Annual Symposium on Switching Circuit and Automata Theory, 1966b.

Page, C.V. Equivalences between probabilistic and deterministic sequential machines. — Information and Control 9, No.5 (1966).

Parshenko, N.Ya. and V.M. Chentsov. On minimization of a stochastic automaton. — Problemy Peredachi Informatsii 5, No.4 (1969). (Russian)

Paz, A.A. Definite and quasi-definite sets of stochastic matrices. — Proc. Amer. Math. Soc. 16 (1965).

Paz, A.A. Some aspects of probabilistic automata. — Information and Control 9, No.1 (1966a).

Paz, A.A. Homomorphisms between stochastic sequential machines and related problems. — Math. Systems Theory 2, No.3 (1966b).

Paz, A.A. Fuzzy star functions, probabilistic automata and their approximation by non-probabilistic automata. — IEEE Conf. Rec. 8th Annual Symposium on Switching and Automata Theory, Austin, Texas, 1967. New York, 1967a.

Paz, A.A. Minimization theorems and technique for sequential stochastic machines. — Information and Control 11, No.2 (1967b).

Paz, A.A. A finite set of $n \times n$ stochastic matrices, generating all n-dimensional probability vectors, whose coordinates have finite binary expansion. — SIAM J. Control 5 (1967c).

Paz, A.A. Regular events in stochastic sequential machines. — IEEE Trans. Electronic Computers C-19, No.5 (1970).

Paz, A.A. and M. Reichaw. Ergodic theorem for sequences of infinite stochastic matrices. — Proc. Cambridge Philos. Soc. 63 (1967).

Perles, M., M.O. Rabin and E. Shamir. The theory of definite automata. — IRE Trans. Electronic Computers EC-12, No.3 (1963).

Podnieks, K.M. On cut-points of some finite stochastic automata. —Avtomat. i Vychisl. Tekhn., No.5 (1970). (Russian)

Podnieks, K.M. On stable expansions of stochastic matrices. — In: Problemy sinteza konechnykh avtomatov. Riga, Zinatne, 1971a. (Russian)

Podnieks, K.M. On a measure of stability for expansions of stochastic matrices. — Avtomat. i Vychisl. Tekhn., No.3 (1971b). (Russian)

Pospelov, D.I. Games and Automata. — Moscow-Leningrad, Energiya, 1966. (Russian)

Pospelov, D.I. Stochastic Automata. — Moscow-Leningrad, Energiya, 1970. (Russian)

Post, E.L. Finite combinatory processes — formulation I. — J. Symbolic Logic 1 (1936).

Rabin, M.O. Probabilistic automata. — Information and Control 6, No.3 (1963).

Rabin, M.O. Real time computation. — Israel J. Math. 1, No.4 (1963b).

Rabin, M.O. Lectures on classical and probabilistic automata. — In: Automata Theory (E.R.Caianiello, Editor). New York, Academic Press, 1966.

Rabin, M.O. and D. Scott. Finite automata and their decision problems. — IBM J. Res. Develop.3 (1959).

Rastrigin, L.A. and K.K Ripa. Statistical search as a stochastic automaton. — Avtomat. i Vychisl. Tekhn., No.1 (1971). (Russian)

Rice, H.G. Recursive real numbers. — Proc. Amer. Math. Soc. 5, No.5 (1954).

Ripa, K.K. Some statistical properties of optimizing automata and of random search. — Avtomat. i Vychisl. Tekhn., No.3 (1970). (Russian)

Salomaa, A. On probabilistic automata with one input letter. — Ann. Univ. Turku. Ser. A I **85** (1965).

Salomaa, A. On m-adic probabilistic automata. — Information and Control **10**, No.2 (1967).

Salomaa, A. On events represented by probabilistic automata of different types. — Canad. J. Math. **20**, No.1 (1968a).

Salomaa, A. On languages accepted by probabilistic and time-variant automata. — In: Proc. 2nd Annual Princeton Conference on Information Science and Systems, 1968b.

Salomaa, A. On finite automata with a time-variant structure. — Information and Control **11**, No.2 (1968c).

Santos, E. Probabilistic Turing machines and computability. — Proc. Amer. Math. Soc. **22**, No.3 (1969).

Shanin, N.A. On constructive linear functionals in constructive Hilbert space. — Z. Math. Logik Grundlagen Math. **2** (1956). (Russian)

Shanin, N.A. On the constructive understanding of mathematical inferences. — Trudy Mat. Inst. Steklov **52** (1958a). (Russian)

Shanin, N.A. On an algorithm for constructive interpretation of mathematical inferences. — Z. Math. Logik Grundlagen Math. **4** (1958b). (Russian)

Shanin, N.A. Some questions of mathematical analysis in the light of constructive logic. — Z. Math. Logik Grundlagen Math. **5** (1959). (Russian)

Shanin, N.A. Constructive real numbers and constructive function spaces. — Trudy Mat. Inst. Steklov **67** (1962). (Russian)

Shanin, N.A. On R.L. Goodstein's recursive analysis and equation calculus. — Introductory article in [Russian translation of] (Goodstein, 1961), Moscow, Nauka, 1970. (Russian)

Sheng, C.L. Threshold logic elements used as a probability transformer. — J. Assoc. Comput. Mach. **12**, No.2 (1965).

Shreider, Yu.A. Learning models and control systems — Supplement to [Russian translation of] (Bush and Mosteller, 1955), Moscow, Nauka, 1962. (Russian)

Skhirtladze, R.L. On synthesis of p-networks from switches with random discrete states. — Soobshch. Akad. Nauk Gruz. SSR **27**, No.5 (1961). (Russian)

Skhirtladze, R.L. On a method for construction of a boolean variable with prescribed probability distribution. — Diskretnyi Analiz, No.7 (1966). (Russian)

Skhirtladze, R.L. and V.V Chavchanidze. On the synthesis problem for discrete stochastic devices.— Soobshch. Akad. Nauk Gruz. SSR **27**, No.5 (1961). (Russian)

Specker, E. Nichtkonstruktiv beweisbare Sätze der Analysis. — J. Symbolic Logic **14**, No.3 (1949).

Stanciulescu, F. and F.A. Oprescu. A mathematical model of finite random automata. — IEEE Trans. Electronic Computers C-**17**, No.1 (1968).

Starke, P.H. Theorie stochastischer Automaten. — Elektronische Informationsverarbeitung und Kybernetik **1**, No. 1/2 (1965).

Starke, P.H. Stochastische Ereignisse und Wortmengen. — Z. Math. Logik Grundlagen Math. **12**, No.1 (1966a).

Starke, P.H. Stochastische Ereignisse und stochastische Operatoren. — Elektronische Informationsverarbeitung und Kybernetik **2**, No.3 (1966b).

Starke, P.H. Theory of stochastic automata. — Kybernetika **2**, No.6 (1966c).

Starke, P.H. Die Reduktion von stochastischen Automaten. — Elektronische Informationsverarbeitung und Kybernetik **4**, No.2 (1968).

Starke, P.H. Über die Minimalisierung von stochastischen Rabin-Automaten. — Elektronische Informationsverarbeitung und Kybernetik **5**, No.3 (1969a).

Starke, P.H. Abstrakte Automaten. — Berlin, VEB Deutscher Verlag der Wissenschaften, 1969b.

Starke, P.H. and H. Thiele. On asynchronous stochastic automata. — Information and Control **17** (1970).

Trakhtenbrot, B.A. Turing computations with logarithmic delay. — Algebra i Logika Seminar **3**, No.4 (1964). (Russian)

Trakhtenbrot, B.A. Optimal computations and frequency effects of Yablonskii. — Algebra i Logika Seminar **4**, No.5 (1965). (Russian)

Trakhtenbrot, B.A. On normalized signaling functions for Turing computations. — Algebra i Logika Seminar 5, No.6 (1966). (Russian)

Trakhtenbrot, B.A. Complexity of Algorithms and Computations. — Novosibirsk, Izdatel'stvo Novosibirskogo Gosudarstvennogo Universiteta, 1967. (Russian)

Trakhtenbrot, B.A. and Ya.M. Barzdin'. Finite Automata. Behavior and Synthesis. — Moscow, Nauka, 1970. (Russian) [English translation: Amsterdam, North-Holland, 1973]

Tseitin, G.S. Algorithmic operators in constructive metric spaces. — Trudy Mat. Inst. Steklov **67** (1962a). (Russian)

Tseitin, G.S. Mean value theorems in constructive analysis. — Trudy Mat. Inst. Steklov **67** (1962b). (Russian)

Tsertsvadze, G.N. Some properties and synthesis methods of stochastic automata. — Avtomat. i Vychisl. Tekhn., No.3 (1963). (Russian)

Tsertsvadze, G.N. Stochastic automata and the problem of designing reliable systems from unreliable elements. — Avtomat. i Telemekh. **25**, Nos.2.4 (1964). (Russian)

Tsertsvadze, G.N. On stochastic automata which are asymptotically optimal in a random medium. — Soobshch. Akad. Nauk Gruz. SSR **43**, No.2 (1966). (Russian)

Tsetlin, M.L. On the behavior of finite automata in random media. — Avtomat. i Telemekh. **22**, No.10 (1961). (Russian)

Tsetlin, M.L. Finite automata and the simulation of elementary modes of behavior. — Uspekhi Mat. Nauk **18**, No.4 (1963a). (Russian)

Tsetlin, M.L. Note on a game played by a finite automaton with a partner using a mixed strategy. — Dokl. Akad. Nauk **149**, No.1 (1963b). (Russian)

Tsetlin, M.L. On the behavior of finite automata in random media. — In: Trudy IV Vsesoyuznogo matematicheskogo s"ezda, II. Leningrad, Nauka, 1964. (Russian)

Tsetlin, M.L. and V.Yu. Krylov. On an automaton which is asymptotically optimal in a random medium. — Avtomat. i Telemekh. **24**, No.9 (1963). (Russian)

Turakainen, P. On nonregular events representable in probabilistic automata with one input letter. — Ann. Univ. Turku. Ser. A I **90** (1966).

Turakainen, P. On probabilistic automata and their generalizations. — Suomalaisen Tiedeakat. Toimituksia Ser. A I, No.429 (1968a).

Turakainen, P. On stochastic languages. — Information and Control **12**, No.4 (1968b).

Turakainen, P. On languages representable in rational probabilistic automata. — Suomalaisen Tiedeakat. Toimituksia Ser. A I, No.439 (1968c).

Turakainen, P. Generalized automata and stochastic languages. — Proc. Amer. Math. Soc. **21**, No.2 (1969a).

Turakainen, P. On time-variant probabilistic automata with monitor. — Turun Yliopiston Julk. Ser. A I, No.130 (1969b).

Turing, A.M. On computable numbers. with an application to the Entscheidungsproblem. — Proc. London Math. Soc. Ser 2 **43** (1936 —1937). Correction. — Ibid. (1937).

Turing, A.M. Computability and 2-definability. — J. Symbolic Logic **2** (1937).

Uspenskii, V.A. Lectures on Computable Functions. — Moscow, Fizmatgiz, 1960. (Russian)

Varshavskii, V.I. and A.M. Gersht. Behavior of continuous automata in random media. — Problemy Peredachi Informatsii **2**, No.3 (1966). (Russian)

Varshavskii, V.I., M.V. Meleshina and M.L. Tsetlin. Behavior of automata in aperiodic random media and problems of synchronization in the presence of noise. — Problemy Peredachi Informatsii **1**, No.1 (1965). (Russian)

Varshavskii, V.I. and I.P. Vorontsova. On the behavior of stochastic automata with variable structure. — Avtomat. i Telemekh. **24**, No.3 (1963). (Russian)

Varshavskii, V.I. and I.P. Vorontsova. Stochastic automata with variable structure. — In: Theory of Finite and Stochastic Automata. Trudy Mezhdunarodnogo Simpoziuma IFAK (Proceedings of International IFAC Symposium). Moscow, Nauka, 1965. (Russian)

Varshavskii, V.I. and I.P. Vorontsova. Use of stochastic automata with variable structure for solution of certain problems of behavior. — In: Samoobuchayushchiesya avtomaticheskie sistemy. Moscow, Nauka, 1966. (Russian)

Varshavskii, V.I. and I.P. Vorontsova. Teams of automata and models of behavior. — In: Samonastraivayushchiesya sistemy. Raspoznavanie obrazov. Releinye ustroistva i konechnye avtomaty. Trudy III Vsesoyuznogo soveshachaniya po avtomaticheskomu upravleniyu. Moscow, Nauka, 1967. (Russian)

Varshavskii, V.I., I.P. Vorontsova and M.L. Tsetlin. Stochastic learning automata. — In: Biologicheskie aspekty kibernetiki. Moscow, Nauka, 1962. (Russian)

Vasariņš, G.E. The problem of recognizing the decomposition length of constructively defined stochastic matrices. — Avtomat. i Vychisl. Tekhn., No.5 (1971). (Russian)

Winkowski, J. A method of realization of Markov chains. — Algorytmy 5, No.9 (1968).

Yasui, T. and S. Yajima. Two-state two-symbol probabilistic automata. — Information and Control **16**, No.3 (1970).

Zaslavskii, I.D. Some properties of constructive real numbers and constructive functions. — Trudy Mat. Inst. Steklov **67** (1962). (Russian)

Zaslavskii, I.D. and G.S. Tseitin. On singular covers and related properties of constructive functions.— Trudy Mat. Inst. Steklov **67** (1962). (Russian)

Zühlke, H. Ersetzbarkeit von stochastischen Automaten. — Wiss. Z. Friedrich-Schiller-Univ. Math.-Naturwiss. Reihe, No.2 (1969).

SUBJECT INDEX